SPREADING OF DISEASES IN PLANTS

RECENT ADVANCES IN PLANT PATHOLOGY SERIES

SPREADING OF DISEASES IN PLANTS

K. M. CHANDNIWALA

ANMOL PUBLICATIONS PVT. LTD.
NEW DELHI-110 002 (INDIA)

ANMOL PUBLICATIONS PVT. LTD.
Regd. Office: 4360/4, Ansari Road, Daryaganj,
New Delhi-110002 (India)
Tel.: 23278000, 23261597, 23286875, 23255577
Fax: 91-11-23280289
Email: anmolpub@gmail.com
Visit us at: www.anmolpublications.com

Branch Office: No. 1015, Ist Main Road, BSK IIIrd Stage
IIIrd Phase, IIIrd Block, Bangalore-560 085 (India)
Tel.: 080-41723429 • Fax: 080-26723604
Email: anmolpublicationsbangalore@gmail.com

Spreading of Diseases in Plants

First Edition, 1995
Reprint, 1998, 2010
ISBN 81-7488-274-X

PRINTED IN INDIA

Printed at Mehra Offset Press, Delhi.

Preface

The present title has been intended primarily for the use of students enrolled in advanced undergraduate and postgraduate courses in plant pathology. The need of a textbook in this field was a pressing one since not only did the students find literature on the diseases and other pathological conditions difficult to obtain and to access, but no reasonably complete treatment of the subject from the student's stand point was to be had. To fill this need is the first objective of this title.

Considering the field as a whole, however, it was realised that in addition to a textbook for classroom use, the subject was also greatly in need of a reference work. The pathologists and the research worker in biology have not had an available guide to all phases of the subject of plant pathology as such for most of the information the research worker has been faced to consult are widely scattered in often almost inaccessible articles or incomplete accounts in larger volumes. Accordingly, the authors, turned their attention to the preparation of the present book, which deals specially with plant pathology, i.e. with the microbial diseases as well as with certain amicrobial deseases, injuries and abnormalities. The task has not been easy one.

Attempts have been made to present uptodate but brief introduction to principles underlying various stages in the development of plant diseases with special emphasis on their control. The language of the book has been painfully kept as simple and lucid style as possible. The subject matter has been interwoven for keeping up interest and curiosity of readers.

There can be no claim to originality except in the manner of treatment and much of the information has been obtained from the books and scientific journals available in different libraries.

Constructive criticism and suggestion for improvement of the book will be thankfully acknowledged.

Author

Contents

1

INTRODUCTION

The spread of parasitic or of virous diseases from plant to plant or from one locality to another involves either the transport of the pathogene or of its spores or of the viruliferous principle. A knowledge of the methods of dissemination is fundamental value in determining and understanding method of control.

Air and Wind Dissemination

The spores of many different species of fungi are prevalent in the air and some of these represent parasitic species. Two general types of fungous spores are adapted to air and wind dissemination: (1) those borne single or in chains on the ends of aerial conidiophores from which they are easily detached, or others that are liberated from the fruiting body as a dry, powdery mass; (2) those which are separated from the spore fruit by some explosive mechanism which forces them into the air, after which they may be carried away by air currents. This prevalence of spores in the air has been demonstrated by exposing plates of culture media to the air and observing the fungi which develop; by filtering air through a trap which retains the spores so that they may be observed under the microscope or identified by planting out in culture media; by the use of spore traps which collect the spores which are set free by an explosive mechanism and carried away by the wind or by convention currents; and by direct microscopic examination or by the use of a light beam. In addition to the *direct* dispersal of spores by air currents, the wind may assist in the *indirect* spread of plant diseases by transporting diseased parts of the host, especially infected leaves or flower parts blown from the diseased tree or later picked up by the wind from the ground litter beneath diseased trees. It may be noted also that wind-blown rain may sometimes transport pathogenic bacteria or fungi which have no special devices for air or wind transport.

Typical examples of fungi adapted to air or wind dissemination of the spores without the use of an explosive mechanism are to be found in many forms of Hyphomycetes such as Penicillium, Aspergillus, Alternaria, Verticillium, etc.; in the conidial stages of the white rutss, downy mildews and powdery mildews; in certain spore stages of smuts and rusts; and in the imperfect stages of some sac fungi or ascomycetes.

Two distinct types of forcible ejection of spores are illustrated: (1) By ascus fruits from which the ascospores are forcibly expelled by the explosion of the asci. Two different types of ascospore expulsion are to be found: (*a*) the successive explosion of the spore sacs from a perithecium until the supply is exhausted, either the entire eight spores shot out at once or in succession like a repeating rifle; and (*b*) the simultaneous explosion of a large number of asci, characteristic of apothecia, giving rise to the phenomenon of "puffing" with the production of visible clouds of spores. Numerous forms of fungi from the Pyrenomycetes including both parasitic and saprophytic species show the first type of forcible expulsion of their ascospores, for example, the powdery mildews, the apple-scab fungus, and the chestnut-tree blight pathogen; while the second type is characteristic of both parasitic and saprophytic species of Discomycetes, for example, the Sclerotinia of wilt, the Sclerotinia of brown rot of stone fruits, the Pseudopeziza of clover and alfalfa leaf spot, the Rhytisma of tar spot of maple, the Coccomyces of cherry leaf spot and others.

(2) By the basidium fruits which provide for the forcible separation of the basidiospores from the basidia. In this case the four spores borne on each basidium are ejected in succession, each spore being thrown off by a jerking process due to the release of hydrostatic pressure. This type of spore discharge is characteristic for the sporidia of the smut fungi, for the sporidia produced from the promycelium of rust fungi, well illustrated by the cedar-rust fungus and other species; and by the various species of fungi producing definite basidia-bearing sporophores or fruiting bodies.

Dissemination by Water

Water may serve for the dissemination of pathogens in two ways: (1) as a medium in which activity motile organisms or spores may swim about; and (2) by the mechanical action of runoff of rain, the flowing of irrigation water, or by stream flow.

Actively motile cells are produced by some bacterial pathogens, the water molds, the chytrids, the white rusts, and some of the downy mildews. Such forms depend wholly or in part upon liquid moisture as a medium in which they may develop and through which they may migrate by the action of their cilia or flagella. Liquid moisture is required for the stomatal or water-pore infection by certain bacteria. The chytrids are primarily aquatic-forms but even forms affecting land plants must rely on liquid soil moisture for the dispersal of their swarm spores, for example in the crown wart of alfalfa, potato wart and clubroot of cabbage and other crucifers. This explains in part why damping-off by *Pythium debaryanum* is favoured by a wet soil. Among the white rusts and downy mildews which develop only on aerial parts, the aquatic habit is still retained in part and liquid moisture is necessary for the formation and dissemination of their swarm spores. This is well illustrated by the late blight of the potato, the white rust of crucifers and the downy mildew of grape.

The dripping of moisture from heavy dews or the runoff from rains may transport nonmotile spores: (1) those which have been lodged on plant parts by wind or other agencies; and (2) those which are residual in origin but are held together by a mucilaginous matrix in which they are embedded. This second condition is illustrated by bacterial exudates as in fire blight of pome fruits and in the olive knot; and by the spore horns or tendrils, extruded from many pycnids or by the gelatinous spore masses formed from acervuli. This production of spores in gelatinous masses is well illustrated by the conidial stage of the chestnut-tree blight fungus and the acervular stage of bean anthracnose. In all of these cases the gelatinous matrix in which the bacteria or the spores are embedded, is dissolved by rains and the separated spores washed down by mechanical action.

Flood water, irrigation water or stream flow may be responsible for the spread of pathogenic forms. Overflow of streams has spread the root and crown rot of celery, while the cultural practice of flooding rice or cranberry fields spreads the disease-producing organisms. Even in row irrigation, the water may carry a load of spores and so increase chances of infection as has been demonstrated for the leaf spot of sugar beets and the late blight of celery. Dissemination by stream flow is well illustrated by the forms of ergot affecting aquatic grasses in which the sclerotia are "fioaters" and so are carried down stream to new hosts.

Dissemination by Insects

Insects may act simply as carriers of infective material adhering to their bodies, or they may harbor the inoculum within their bodies, and in certain cases make the inoculation by feeding wounds, either by chewing or sucking mouth parts. Both bacterial and fungous diseases are disseminated by insects but they are of outstanding importance as the vectors of virous diseases. In relation to insect dissemination the following groups of plant diseases may be recognized:

1. Organism carried externally and infection accidental, without direct inoculation by the insect. Examples: fire blight of pear and other hosts and bacteriosis of walnut; rusts and smuts of numerous hosts, bitter rot of apple, chestnut-tree blight, late blight of potato, Cercospora leaf spot of beet, cotton anthracnose and many others.

2. Organism or a virus carried externally and infection by mechanical transfer through feeding by the insect carrier. Examples: fire blight of pear and apple, bacterial wilt of Solanaceous plants, black rot of crucifers and some other bacterial diseases; downy mildew of Lima bean, brown rot of stone fruits, bud rot of carnations, blackleg of cabbage and many other fungous diseases; and certain virous diseases carried by biting insects.

3. Organism not carried by insects, but infection through wounds made by insects. Examples: bacterial leaf spot of cotton and of sugar beet; onion blight powdery mildew of grasses chestnut-tree blight and others.

4. Organism carried internally by insects, but without any specific biological relation, that is the carriage is purely mechanical. In many cases these same organisms are also carried externally. In this mechanical internal transmission the spores remain viable after passage through the alimentary tract but do not increase in number. This behavior is illustrated by the spores of the Sphacelia stage of ergot which have been shown to survive passage through the digestive canal of small arthropods and gastropods.

5. Organism or the infective principle carried internally in the body of the insect carrier and undergoing a multiplication of increase sometimes only after a definite incubation period. This method of transmission has been proved for certain bacterial diseases but is or outstanding importance for many virous diseases. Among the bacterial diseases falling in this class are the bacterial wilt (*Bacillus tracheiphilus*) of cucurbits carried by the striped and 12-spotted cucumber beetles, and the olive knot (*Pseudomonas savastanoi*) carried by the olive fly. The majority of virous diseases, especially those transmitted by sucking insects belong in this group.

Dissemination by other Animal Life

The other forms of animal life credited with the dissemination of diseases are nematodes, slugs or snails, birds and wild or domestic mammals. Specific cases are the spread of bacterial brown rot of potatoes by nematodes, black rot of cabbage and other crucifers by slugs, the chestnut-tree blight by sap-suckers which drill into the bark, certain wood-rotting fungi by mice and squirrels, and potato scab and wart by manure from horses, cows, or pigs fed on the diseased tubers. It has also been pointed out that the spores of certain parasitic forms mingled with the soil may be carried from contaminated fields to disease-free fields on the feet of domestic animals. Such transport has been reported for potato wart, onion mildew, and soil-borne pathogens of cabbage.

Dissemination by Seed

As considered in this relation, seed will include true seed, fruits functioning as seed, and also vegetative reproductive structures, $uch as tubers, fleshy roots, bulbs, corms and rhizomes. A pathogen may be carried *on* or *within* these reproductive bodies or, in some cases, as separate bodies mingled with the seed.

When the infective material is carried only on the surface or mingled with the true seed, the contamination usually occurs during the threshing or separation of the seed. The presence of surface-borne spores can be demonstrated by washing the seed in water, followed by a microscopic examination of the sediment, either before or after concentration by centrifuging. Some of the recorded cases of surface-borne pathogens on seed are the black rot (*Pseudomonas campestris*) of cabbage, the wilt (*Fusarium lini*) of flax, bunt (*Tillctia spp.*) of wheat, covered smut (*Ustilago levis*) of oats, covered smut (*Ustilago hordci*) of barley, stalk smut (*Urocystis occulta*) of rye, millet smut (*Ustilago crameri*) and sorghum and broomcorn smut (*Sphacelotheca sorghi*). Two notable cases of inclusions mingled with the seed are afforded by ergots of rye and other cereals and the seeds of clover and alfalfa dodder that are mixed with the seed during threshing operations.

Some of the diseases known to be carried by the organism within true seeds are the bacterial blight (*Pseudomonas phaseoli*) of beans, the Phytophthora disease (*Phytophthora phaseoli*) of Lima beans, the anthracnose (*Colletotrichum lindemathianum*) of navy and string beans and many other diseases caused by imperfect fungi. Fruits that function as seeds may harbor the parasite within the seed coats or within the embryo. This has been shown to be true for the anthracnose (*Col-*

letotrichum cereale) and the scab (*Fusarium roseum*) of wheat, the spot blotch (*Helminthosporium sativum*) of barley and the loose smuts (*Ustilago tritici* and *U. nuda*) of wheat and barley. Among the fruits other than the caryopses of cereals functioning as seeds which are known to harbor the pathogene are beet seeds containing *Phoma betae*, celery seed (mericarps) containing the pycnidia of *Septoria petroseleni* and the seed (achenes) of oyster plant containing mycelium and oöspores of white rust (*Albugo tragopogonis*).

Most of the virous diseases are not carried by the true seeds, but seed carriage is general for some legume mosaics and has been demonstrated for the mosaic of lettuce, the mosaic of petunia and the mosaic of the wild cucumber.

Many causal agents of disease, bacteria, fungi, nematodes, or viruses, are carried on the surface or within the vegetative reproductive structures or storage organs such as roots, tubers, bulbs, corms or rhizomes. The sweet potato and other root crops, the Irish potato, bulbs of the onion, and the corms, bulbs and rhizomes of flowering plants are frequently the carriers of numerous diseases.

Dissemination by Propagating Stock

Cuttings herbaceous seedlings, scions used for grafting, and nursery stock may act as carriers of both parasitic and virous diseases. The propagating stock may act simply as a carrier of the spores of a pathogen; incipient or established infections of a pathogen may be present; viruses may be present and still latent; or visible symptoms may be evident. Infections developed within the seed bed may be carried to the field by some of the transplants. This has been shown to be true for several sweet-potato diseases (Fusarium stem rot, dry rot, and foot rot) at least three different cabbage diseases, and especially for the virous diseases of tobacco.

For fruit, forest or shade trees, or shrubs any diseases which affect the roots or woody stems may be introduced with the young trees. Notable examples are crown gall (*Pseudomonas tumefaciens*), plum black knot (*Dibotryon morbosum*) and chestnut tree blight (*Endothia parasitica*), the latter probably first brought to America on nursery stock imported from China. It is believed that citrus canker and the white-pine blister rust were also introduced with imported nursery stock, the former from Japan and the latter from Europe.

Dissemination of Crude or Commercial Plant Products

These include such products as fruits, vegetables, grain, hay, straw, packing materials and forest products, that are not used for propagative purposes. Examples may be found in such diseases as scab, black rot and bitter rot of apples, the brown rot of stone fruits, the scab, wart and powdery scab of potatoes and cabbage and other vegetable diseases. Transport of a few other diseases may be noted from recorded cases: gooseberry mildew on packing material; cereal diseases in grain, hay or straw; cereal smuts in meal, bran, flour or contaminated bags; and chestnut-tree blight on logs, poles, posts, ties and tan bark.

Dissemination with Soil, Litter, Compost or Manure

By the transport of these materials parasitic organisms may be introduced into new environments. The parasites introduced may be either organisms that normally inhabit the soil or they may be accidental inclusions. Damping-off fungi may be brought into the greenhouse with contaminated soil, Sclerotinia wilt of lettuce, tomatoes and other vegetables may be introduced by sclerotia mingled with the soil or with plant refuse, or the use of alfalfa soil for inoculating new fields with legume bacteria may introduce soil-inhabiting pathogens. Numerous other cases of soil contamination by soil carriage or litter have been recorded. including cabbage yellows, flax wilt, carnation wilt, Mycogone disease of mushrooms, black rot, charcoal and stem rot of sweet potato, Rhizoctonia and Fusarium diseases of Irish potato, and the smut of onion.

Forms that are not primarily soil dwellers, may spend a part of their life cycle in the soil or soil refuse. The oöspores of onion mildew (*Peronospora schleideniana*) may be carried with onion leaf litter, cotton anthracnose (*Glomerella gossypii*) on cotton hulls, Ascochyta blight (*Mycosphaerella pinodes*) of peas in pea straw, the leaf spot (*Cercospora beticola*) of beet in the old tops, and, when these are left *in situ* or scattered over new areas, they may continue the diseases. In none of these cases is there any multiplication of the organism in the soil, but, in the case of corn smut, the spores germinate in the soil or in the compost heap and give rise to numerous secondary spores which may cause a heavy soil contamination.

Dissemination by other Agricultural Practices

Plant diseases may be spread by the ordinary operations of transplanting seedlings, by watering, by spraying, by picking or harvesting operations, and by threshing or ginning. A few specific examples may be cited.

Blackleg of potatoes may sometimes be spread when cutting seed tubers, but experimental studies have shown this method to be less important than formerly reported; fire blight of pear and apples in pruning; late blight of potatoes during digging and handling; mosaic of tobacco during transplanting or when topping in the field; tomato streak, a virous disease, by tobacco users during the pruning operations; bean anthracnose during cultivation and picking; brown rot of lemons by washing previous to curing; blue mold and other rots of apples by the solutions used for spray-residue removal; covered smuts of cereals by the threshing operation which not only contaminates the sound grain but scatters the smut spores over surrounding fields; and cotton anthracnose by the ginning of the cotton.

2

INTER-RELATIONSHIP BETWEEN SOIL AND PATHOGEN

The interaction of soil microfloras on plant growth exhibits considerable plasticity. For instance, certain microorganisms, particularly bacteria, enter into a commensalism with roots and help certain leguminous plants in acquiring the all important nitrogen for their metabolic needs. The peculiar symbiotic relationship of the endotrophic and ectotrophic mycorrhizal fungi with root systems of certain forest trees is another facet of the problem. Heterotrophy among mycorrhizal fungi became evident with the recent discovery of the B vitamins. Radiophosphorus was shown to be absorbed by endotrophic mycorrhizal fungi and transferred to the host tissue.

Nevertheless, there are other relationships between plants and microorganisms of the soil that are not looked upon with favor since their destruction of crops can deprive man of one of his primary needs. Microorganisms may compete with plant roots for essential energy substrates for development of enzyme systems and primary cell functions.

It therefore becomes imperative to study interactions between pathogens and other microflora of the soil and also between pathogens and the host root systems. Around this has grown the new concept of rhizosphere microfloras and their repercussions on host-pathogen relationships. This shift in emphasis is fortified by the discovery that many amino acids, sugars, organic, and inorganic substances are exuded from living root systems and that antibiotics and other metabolites are constantly being produced *in situ* in soils. The uptake of these complex microbial metabolic products is no longer a matter for speculation ad there seems every reason to believe that these interfere *in vivo* with ionic balance and lead to consequent enzyme dysfunction. Clearly, therefore, in the study of soil-borne root infections we have to consider the varied and complex interactions between pathogen, soil, and host.

I. PATHOGENS IN SOIL

Soil microbiology as a science has come into its own and shown that there is a dense population of varied microbes in the soil. This microbial population includes several hundreds of forms of fungi, many bacteria as well as actinomycetes. It includes simple saprophytes and plant pathogens. The pathogenic forms are capable of infecting roots of suitable hosts and cause disease. The very stable tobacco mosaic virus and certain other viruses such as wheat mosaic, lettuce big vein, and tobacco necrosis viruses, which lose their infectivity much more rapidly, are soil-borne.

Soil-borne plant pathogens differ in the type of disease they produce and these root diseases may be classified on the basis of the nature of the tissues attacked and the effects produced.

Parenchyma diseases are those in which the parenchyma alone may be the center of attack and may be disintegrated. These diseases may be caused either by soil bacteria, or fungi. Soft rot of carrots, for example, is caused by *Bacterium carotovorum* and the pathological effects produced result from the separation of the individual cells of the parenchyma due to enzyme action whereby the middle lamella between adjacent cells is dissolved. On the other hand, several species of fungi belonging to the genera *Pythium, Phytophthora, Rhizoctonia*, etc. cause root rots, foot rots, stem rots, and damping-off of seedlings. In all these cases, rotting may be due to death of cells or to the dissolution of the middle lamellae. The early work of de Bary and of Marshall Ward indicated that, following penetration of host tissue, *Sclerotinia* and *Botrytis* secreted enzymes which diffused in advance of the pathogen and led to the breakdown of parts of the cell membrane.

Vascular diseases are those in which the vascular tissues, particularly the xylem, may be invaded and blocked resulting in wilting. These may also be caused by bacteria or fungi. As examples of bacterial vascular wilt may be cited the gumming disease of sugar cane due to *Xanthomonas vasculorum* and bacterial wilt of maize due to *Xanthomonas stewartii*. Vascular wilts due to fungi are many and are caused by species of *Fusarium, Verticillium, Valsa*, etc. The symptoms usually include various degrees of chlorosis, vascular discoloration, stunting, and wilt. Infection takes place through root systems and the pathogens are normally confined to the vascular elements except during the final stages of the disease when the plant tissues are dead and the pathogens grow out from the vascular elements. Cotton wilt due to *Fusarium vasinfectum* and tomato wilt due to *F. oxysporum* f. *lycopersici* are good examples.

There are then what may be called *systemic diseases* in which both parenchyma and vascular tissues may be invaded. For instance, in the case of brown rot of solanaceous plants due to *Pseudomonas solanacearum*, the symptoms are those of a general wilt followed by collapse. In most affected plants, a brown stain is invariably formed in the xylem. If infected stems are cut across, the bacteria ooze out in drops. In contrast to the vascular wilts, the pith and the cortex are invaded later. Distinct cavities filled with bacteria are formed in the pith region and the cortex may undergo disintegration.

There are also many root infections which lead to *hyperplastic diseases*. Clubroot of crucifers due to *Plasmodiophora brassicae*, potato wart due to *Synchytrium endobioticum*, and crown wart of alfalfa due to *Physoderma alfalfae* are well known examples of hyperplastic diseases of fungal origin. Hairy root and crown gall of plants, on the other hand, are induced by bacteria infecting root systems. What is noteworthy about these galls is the capacity of unlimited and unrestrained proliferation acquired by some of their cells in the absence of the crown gall bacterium.

II. PATHOGENS AND ROOT INFECTION

A. Facultative and Obligate Parasites—Host Range

In the etiology of root disease, infection of root systems takes place either by an obligate parasite or a facultative one, or sometimes even by quite a different agent such as a virus. Problems relating to obligate and facultative parasites have been admirably discussed by Brooks (1948) and Brown (1948) and it would be sufficient here to consider the question of obligate and facultative parasitism insofar as it is peculiar to root-infecting pathogens. Garrett (1956) has built up an evolutionary sequence beginning with the obligate soil saprophytes lacking the ability to parasitize living plants and ending with what he terms "ecologically obligate parasites", whose relationship with the hosts is one of symbiosis.

The intermediate links in this evolutionary chain are (1) the primitive parasites that destroy seedlings and juvenile tissues and can live saprophysically; (2) the less primitive parasites which rapidly destroy plant tissues and are less restricted by the host than (1); (3) the specialized parasites that cause less disorganization of the host tissues and do not exhibit saprophytic activity other than passive survival in tissues invaded as parasites.

Although Garrett's classification refers to root-infecting fungi, it should be applicable to other root-infecting pathogens also. However, in our view, this evolutionary sequence does not necessarily imply an inverse relationship between parasitic specialization and saprophytic ability such as has been underlined by Garrett throughout his treatise on the biology of these pathogens. A facultative parasite, for instance, in having acquired the ability to parasitize a host, certainly need not be considered to have lost its saprophytic ability even in a small measure. In other words, it is not less of a saprophyte than the so-called obligate saprophyte because it has acquired the ability to parasitize a host plant.

That this is so becomes obvious from a comparative analysis of the competitive saprophytic ability and parasitic specialization seen among root-infecting fungi. For instance, comparison may be made between *Fusarium vasinfectum*, the cotton wilt pathogen, and *Sclerotium cepivorum* which causes white rot of onions are related hosts. The former causes a typical vascular wilt and, in its relationship with the host, hardly causes any decay or necrosis, but normally invades the xylem elements whence it produces metabolites (vivotoxins) whose activity on host tissues leads to wilting. *Sclerotium cepivorum*, on the other hand, causes a rot of the host tissues, presumably by enzymatic action. Insofar as *Fusarium vasinfectum* causes little necrosis of host tissues and tends toward a somewhat balanced relationship with its host, it may be considered more specialized than *Sclerotium cepivorum*. When, however, the competitive saprophytic ability of these two pathogens is compared, it is found that they are closely similar. That *Fusarium vasinfectum* makes little free mycelial growth in unsterilized soils, but is capable of long periods of survival therein as a colonizer on dead plant remains, is now an established fact. *Sclerotium cepivorum* behaves in exactly the same way, as shown by the recent work of Scott (1956). Thus, *Fusarium vasinfectum*, notwithstanding its parasitic specialization, possesses as much competitive saprophytic ability as that shown by *Sclerotium cepivorum* and, indeed, both may be grouped as soil inhabitants, as defined by Garrett (1956). On the other hand, *Fusarium udum*, which causes a vascular wilt of *Cajanus cajan* similar to cotton wilt, appears to possess much less competitive saprophytic ability than *Fusarium vas-infectum*, although the parasitism of these two pathogens is practically of the same order of specialization. Indeed, in the case of *F. udum* it has been shown that it can persist in soils only as a colonizer on host tissues which it originally invaded as a parasite, a behavior strikingly reminiscent of that of *Ophiobolus graminis*. These examples clearly show that little support can be accorded to the view that specialization of parasitism brings along with it loss in competitive saprophytic ability. While this is so in some cases, it certainly cannot be a general rule and cannot, therefore, form the backbone of any logical arrangement of behavior patterns of these organisms.

The concept of "ecologically obligate parasites" proposed to take in the mycorrhizal fungi, again suggests that the mycorrhizal habit is the end product of specialized parasitism. This may be so in some cases but, as pointed out by Bawden (1957), "there seems no need to assume that such conditions only arise from a long period of evolution, or that parasites must always evolve from virulence through avirulence to symbiosis."

The subject of host range necessarily requires treatment in any consideration of parasitic behavior of root-infecting pathogens. Here again an extraordinary range in variability is met with, with many pathogens showing preferential pathogenicity on a few host species and others being omnivorous. Between these two extreme type intermediate forms also occur. Parasitic specialization leading to balanced or obligate parasitism need not carry with it restrictions in the choice of hosts, although this does occur in some cases. Conversely, some facultative parasites, although undoubtedly less specialized than the balanced ones, may exhibit a high de3gree of specialization in host choice. Consider, for instance, *Plasmodiophora brassicae* which causes clubroot of crucifers. For a long time, this obligate parasite was considered to be capable of infecting only cabbage and other crucifers, but the recent work of Webb and of Macfarlane has removed this misconception. From the work of these investigators it is clear that this fungus can infect root systems of noncruciferous plants such as *Tropaeolum majus, Reseda odorata, Papaver rhoeas, Agrostis alba* var. *stolonifera, Dactylis glomerata*, and *Lolium perenne*. In none of these, however, is the fungus known to produce clubroot symptoms following infection. Root hair infection of seedlings of *Matthiola incana, M. bicornis*, and of *Lepidium sativum* has also been observed and in *L. sativum,* infection was followed by clubroot symptoms. Further, among the Cruciferae, the pathogen has a wide host range. Similarly, *Synchytrium endobioticum* is another obligate parasite which infects, besides potato and tomato, *Solanum nigrum, S. dulcamara*, and other species of *Solanum, Hyoscyamus niger* and *Nicandra physaloides*. The host spectrum of this fungus is thus limited to the Solanaceae. *Spongospora subterranea*, which causes powdery scab of potatoes, likewise infects a few other hosts belonging to the Solanaceae. A further restriction in host range is met with in the case of another obligate root parasite, *Physoderma alfalfae*, which infects only common alfalfa, *Medicago sativa* and *M. falcata*. It would appear that none of the root-infecting obligate parasites shows preference to less than two host species.

The greater number of root-infecting pathogens, no doubt, are of the facultative type and among these also we do find a narrow or a wide host range. Species of the genera *Pythium, Sclerotinia, Rhizoctonia, Armillaria, Fomes, Ganoderma, Fusarium*, and *Verticillium*, to cite only a few examples, are omnivorous and can infect an appallingly wide range of hosts which may be closely related or not related at all. Other facultative parasites appear to have a host spectrum limited to related plants belonging to a single family or sometimes to a single genus. Thus, *Sclerotium cepivorum, Botrytis allii*, and *Colletotrichum circinans* can infect only species of the genus *Allium*, and *Fusarium oxysporum* f. *lycopersici* attacks only two species of *Lycopersicon*. Examples may also be cited here from among soil-borne bacterial pathogens. *Corynebacterium fascians*, which causes leafy gall and fasciation in chrysanthemum, dahlias, etc., is apparently specialized in its parasitism since it causes hyperplastic disease, but is able to infect plants belonging to different families. *Pseudomonas solanacearum* causes brown rot of several Solanaceae, and *bacterium tumefaciens* infects a large number of related and unrelated host species. On the other hand *Xanthomonas hyacinthi* infects only a single species. *Hyacinthus orientalis*, and *Xanthomonas vasculorum* infects only sugar cane.

The examples just cited include root-infecting pathogens of varying degrees of specialization in regard to the type of disease caused. Progressive limitation in host range does not necessarily accompany progressive specialization in parasitism of root-infecting pathogens. These two are obviously independent tendencies which have found expression to varying, but not necessarily equal, degree in different root parasites.

B. Infection: Prepenetration and Penetration Phases—Physical, Chemical, and Physiological Aspects

Root infection is an essential condition for root disease. The process of infection in root disease has been studied by a number of workers and the principles underlying root infection are in general much the same as those that hold for aerial infections. Spores of pathogenic fungi or inocula of other disease causing agents occur in soils, often in great abundance, and in the case of sporulating forms there are obviously a number of factors which would affect their germination. Indeed, in the case of all pathogens, infection depends on various factors: (*a*) viability; (*b*) conditions suitable for germination and penetration. In any given soil, spore germination depends largely on a suitable combination of moisture content, temperature, oxygen supply, pH and, in some cases, the presence of suitable nutrilites.

Many years ago, Brown (1948) showed that there is a certain amount of "exosmosis" of materials from host tissues, leading to a chemotropic stimulus for spore germination and infection. This postulate is particularly true of soil-borne pathogens since it is known that root exudates from living plant roots may influence spore germination and root infection considerably. For instance, Noble (1924) showed that germination of spores of *Urocystic tritici* was stimulated by roots of nonsusceptible plants, e.g., pea, bean, and rye, ând also by traces of benzaldehyde, salicylaldehyde, acetone, and butyric acid. Macfarlane's work (1952) indicates that similar stimulation of resting spore germination in *Plasmodiophora brassicae* is brought about by several cruciferous and noncruciferous species resistant to clubroot. In the same way, *Datura stramonium* has been reported to stimulate germination of spore balls of *Spongospora subterranea*, without the subsequent development of the disease in the root system. The potato plant is known to secrete into the soil substances tending to break the dormancy of spores of the potato wart fungus, *Synchytrium endobioticum*. Varieties of flax resistant to wilt caused by *Fusarium oxysporum* f. *lini* are known to excrete hydrocyanic acid through their root systems and this was shown to depress the growth of the pathogenic *Fusarium* and also of *Helminthosporium*. Further, hydrocyanic acid was also shown to encourage the growth of saprophytic fungi such as *Trichoderma viride* which are antagonistic to plant pathogenic forms. Root excretions of susceptible flax varieties do not appear to contain appreciable amounts of hydrocyanic acid. In the same way, Buxton (1957) found that exudates from the roots of pea varieties stimulate the germination of spores of races of *Fusarium oxysporum* f. *pisi* to which the varieties are susceptible and inhibit that of races to which they are resistant. He also found that, when a variety resistant to the race present in the soil was grown, wilt incidence of a susceptible variety grown in that soil subsequently decreased.

These observations are of great significance in understanding the biology of root infections since they indicate that spore germination around the rhizosphere and subsequent infection would be greatly influenced by exudations from susceptible and nonsusceptible root systems. Where stimulation of spore germination is achieved with the aid of nonsusceptible plants, possibilities of reducing the effective survival of spores in soil and bringing down the inoculum potential to an innocuous level have been envisaged. However, little is known about the stability of these exudates in soils and it is probably not likely that they remain effective to a sufficient extent in space and time to be active in breaking spore dormancy.

The infection process itself deserves to be discussed here in some detail since we do find among the root-infecting fungi distinctly different patterns of behavior. De Bary's hypothesis of "killing in advance of penetration" proposed to explain infection by facultative parasites was challenged and disproved by Brown (1948) who worked with *Botrytis cinerea* and presented evidence showing that the host cells are not killed by this facultative parasite until after it penetrates the cuticle of the host. According to Brown, direct penetration through cutinized walls is believed to be entirely by mechanical pressure, since no cutin-dissolving enzyme has been demonstrated in fungi. The root-infecting fungi normally encounter a host substratum devoid of cutin. Penetration of such noncutinized walls may be either by mechanical pressure or by the dissolving action of enzymes secreted by the fungus, or both. Many pathogens enter the host at the points of emergence of lateral rootlets and several infect through wounds. The so-called "wound parasites" are those which cannot infect by direct penetration but only through wounds. Such wounds may be the result of root injury due to various causes, including injuries due to animals.

Finally, mention must be made of root infection by plant viruses. There is evidence that such infection does occur, notwithstanding the conflicting reports published by different workers. For tobacco mosaic virus it has been claimed repeatedly that the soil is a common source of primary infection; on the other hand, it has also been shown that root infection does not take place commonly even if the roots are deliberately wounded. In the case of tobacco necrosis viruses, the roots get easily infected, but there is little movement of the viruses from the roots to the tops and the tops show no symptoms. This is surprising in view of the fact that when the leaves are inoculated with the viruses, they spread to all plant parts including the roots. That this reluctance to invade the tops from the roots is not common to all viruses is shown by the work of Roberts (1946) who found that potato virus X can spread from infected to healthy plants which are in contact only below ground, and tomato plants infected through their roots showed leaf symptoms.

Granting that parasitism and specialization are later products of an evolutionary tendency, we have then among the root-infecting pathogens infection behavior patterns from the simple to the highly specialized. The forms that "kill in advance of penetration" should possibly find a place low in this scale of evolution, followed by the so-called "wound parasites." The forms that show the ability to penetrate mechanically the host surface are evidently more specialized. The peak of this tendency is found in some pathogens such as *Ophiobolus graminis* causing take-all of wheat and *Phiulophora radicicola* attacking maize roots where the fungi spread extensively along the living root systems of the hosts mainly by brown runner hyphae and produce, in addition, fine infection hyphae which penetrate the host tissue at many points. Finally, it is a point of more than ordinary interest that most of the obligate root parasites are capable of direct penetration of at least younger root tissue.

C. Factors Concerned in Pathogenicity

1. *The Pathogen: Inoculum Potential; Physiological Requisites for Pathogenicity and Disease Development—Enzymes, Vivotoxins, Antibiotics, etc.* Since infection is the first step in pathogenesis, presence of the pathogen is also implied as a prerequisite. Usually, mere presence of the pathogen alone is also insufficient for successful infection and pathogenicity; inoculum of the pathogen will have to be present in quantity. Inoculum potential, then, is the effective concentration of inoculum per unit area of root surface needed for pathogenicity. The term thus connotes not only presence but also effectiveness of the inoculum. Both the concentration and effectiveness of inocula in soils will depend on the many factors. Depending on the conditions under which the host and the pathogen interack, and also the nature of the host and the pathogen a minimum concentration of infecting units may be needed for successful infection. Following Gäumann (1950), this may be termed the numerical

threshold in infection. Where, as in susceptible hosts and virulent pathogens, the chances of infection are high, the numerical threshold is low. In the case of root diseases in general, a single spore may be ineffective. In the case of potato wart, for example, 200 resting sporangia of *Synchytrium endobioticum* per gram of soil are necessary to cause disease. Similarly, in the case of snow mold of wheat due to *Fusarium nivale*, at least 10,000 conidia per milliliter of inoculum are needed; in tomato wilt due to *F. oxysporum* f. *lycopersici*, even with a virulent strain, 700,000 conidia per milliliter of inoculum are essential for pathogenicity. The further need of a food base providing nutrients is seen in many root-infecting pathogens such as *Fomes lignosus, Armillaria mellea, Xylaria mali, Phymatotrichum omnivorum*, and *Rhizoctonia crocorum*.

Many other factors influence pathogenicity, among which are enzymes, vivotoxins, and antibiotics. The essentiality of enzymes for living systems is generally appreciated. It did not, however, fall to the lot of plant pathologists to unravel the mysteries of enzyme chemistry of phytopathogenic fungi and bacteria until the early twenties when Brown (1915) initiated his studies on enzyme systems of fungal facultative parasites and, indeed, he has recently reviewed the entire field in a masterly fashion.

From the plant pathological point of view, pectolytic enzymes that are known to attack cell wall constituents and their production *in vitro*, and the ultimate correlation of this knowledge of *in vivo* functional mechanisms have found much favor and many active preparations from fungi, such as *Botrytis cinerea, Pythium* spp., and many soft rotting bacteria have been made. There is, however, no parallelism possible between the pathogenic potentialities of the organism and its *in vitro* enzyme production.

Wherever there is evidence of enzymatic action on the cell wall it is generally conceded that there would appear no need to postulate the presence of a killing agent. It is true that there is evidence for a number of microbial metabolites injuring host protoplat, but not in all cases have there been justification and experimental proof for their production *in vivo* in sufficient concentration to have caused the injury. There seems little doubt that substances of large molecular sizes can be dislodged or liberated by enzymatic action on cell pectic substrates and these can create a situation in vessels and possibly in the pores of pit membranes resulting in clogging and consequent internal resistance to water conduction.

Much applied work on wilt toxins and application of this fundamental knowledge of pectolytic enzymes has been reported in recent years. Reviews on the subject by Dimond (1955) and Walker and Stahmann (1955) have admirably summarized these facts.

It is this newer concept born of fundamental studies on pectolytic enzymes of the cell that has to be considered next under the head vivotoxin. The new term "vivotoxin" has been defined as a substance produced by the pathogen or its host, or both, which operates in the causation of disease but is not in itself the primary agent. Lycomarasmin produced *in vitro* by *Fusarium oxysporum* f. *lycopersici* does not seem to answer the test and is probably not a vivotoxin, but fusaric acid (produced *in vitro* by a number of fusaria: *F. vasinfectum, F. heterosporum, F. oxysporum* f. *lycopersici, F. orthoceras, Gibberella fujikuroi*) is now regarded as a certainty in opening the new list of vivotoxins along with ethylene producd in *Fusarium* cultures Relatively little, therefore, is known of this group of toxins and it would be necessary to analyze critically newer toxins of microbial origin that affect plants and assign them by grouping together metabolites with specificity of action in the causation of disease syndrome.

Antibiotics of interest to plant growth and disease production have been in the limelight for well over a decade and this subject has been reviewed more frequently than any other branch of plant pathology. Much work has been done on the stability of antibiotics in soils when added from external sources, and positive proof of their remaining potent for long periods under varying microbial antagonism has been obtained. Likewise, the production of antibiotics in soils *in situ* by common soil organisms not only has been provide to occur in many types of soils but it is now considered to be of fairly widespread occurrence. Furthermore, the movement of quite a number of antibiotics from soils into plants, through their roots into their leaves, with considerable rapidity, depending on the metabolic state of the plant and its environment, is now an established fact. Fusaric acid, a pyridine-carboxylic acid, is an example of a vivotoxin and, fusaric acid, like lycomarasmin, has metal binding properties. Much work done with *F. vasinfectum* and neat fusaric acid on the cotton plant has recently been summarized. Some of the highlights may be mentioned here. The addition of zinc to *in vitro* cultures of *F. vasinfectum* increases the output of fusaric acid, the optimum being 0.24 mg. per liter and levels higher than this inhibit the production of this toxin. Possibly, nonprotein source of nitrogen in the host is more conducive to fusaric acid output, and diploid susceptible varieties of cotton have less protein and more of nonprotein nitrogen. Cystine occurring in higher quantities in the resistant tetraploid cotton variety tested appears to be a limiting factor to wilt and presumably to fusaric acid output *in vivo*. Curiously enough, soil amendments with zinc—and growing susceptible plant in them—result in the liberation of cystine even in these plants which appears to confer resistance to *F. vasinfectum* in infested soils, and the susceptible variety virtually behaves like the resistant one. Growing susceptible diploid cotton plants in inoculated soils at temperatures of 32°, 35.0°, and 37.5°C resulted in an apparent recovery or masking of symptoms at 37.5°, whereas at the two lower temperatures wilt symptoms were prominent. Such recovered plants, however, showed higher quantities of fusaric acid and cystine than low temperature ones, indicating that this excess cystine chelated with some heavy metal(s) *in vivo*, thus rendering them unavailable for fusaric acid potentiation. Spectrochemical analysis of the ashes of susceptible diploid and resistant tetraploid cottons, healthy and infected, showed depletion of the key element potassium in tissues of infected susceptible plants, whereas calcium, magnesium, and other elements were on the increase. However, the over-all picture was one of increase in metallic accumulation as shown by greater conductivity in the diseased susceptible plants. Further, there was a basic difference in the pattern of ionic absorption between the diploid and the tetraploid plants, the former taking up much higher quantities of cations than the latter. It appears that further studies on the plasma membranes and the gene controlled mechanisms of these genetic materials and its consequent repercussions on *in vivo* chelations with vivotoxins may add considerably to our knowledge of the behavior pattern of host plants under toxemia.

Pathogenesis is a connected sequence of events involving extraneous sources of energy substrates in the soil such as cellulose, essential minerals, vitamins, sugars, protein breakdown products including a multiplicity of amino acids, and, in fact, complex fractions such as root exudations and products of microbial synthesis such as the enzymes, vivotoxins, and antibiotics. These should be assembled before us if we are to understand every stage leading to reversible and irreversible *in vivo* changes in plants.

Recent investigations by Paquin and Waygood (1957) seem to lead one on to a new approach. They have indicated the presence of an active cyclophorase system capable of oxidizing the principal acids of the Krebs cycle in mitochondria from hypocotyls and cotyledons of tomato seedlings. Lycomarasmin and fusaric acid, produced by *Fusarium oxy-*

sporum f. *lycopersici*, inhibit the succinoxidase and cytochrome oxidase activity of the mitochondria at a concentration of 10^{-2} *M*. The inhibition was overcome by the addition of catalytic amounts of cytochrome c. No effect was observed on succinic dehydrogenase. The toxins seem to inhibit enzyme activity by affecting the integrity of the mitochondria allowing diffusion of cytochrome c from active sites. These authors support the idea that lycomarasmin and fusaric acid could possibly affect host metabolism by removing cytochrome c from the active sites, thus causing disturbances in linked-enzyme systems rather than by destroying the semipermeability of the plasma membrane. They also categorically state that lycomarasmin and fusaric acid are not involved in the wilting of tomatoes because concentrations of toxins required for the complete inhibition of the succinoxidase activity of tomato mitochondria appear to be too high for these toxins to be specific inhibitors and, further, the lycomarasmin inhibition of the succinoxidase activity was not increased by the presence of free iron ($FeCl_3$). They, however, admit that the observation of the presence of free iron leading to potentiation of lycomarasmin is a valid one. In other words, iron chelation *in vivo*, already referred to, is not ruled out.

The enzyme story is a sound enough argument for the initiation of tissue disintegration and especially middle lamellae of cells but may not itself be adequate to explain the whole sequence of events. The rapidity of movement of the vivotoxins must be studied more vigorously. If cytochrome c disturbance is not to be considered to explain fully the alleged loss of semipermeability of the plasma membrane as indicated by ionic derangement, further experimental work has to be undertaken to find out alternative explanations for this rapid ionic imbalance. A wider range of genetic plant material should be used inasmuch as the pattern of movement of ions appears to be very different in diploid and tetraploid cotton plants. One shrinks from suggesting the inclusion of polyploids, allotetraploids, and other genetic materials.

2. *The Host: Resistance or Susceptibility*. That pathogenicity will also depend on the innate nature of the host is too well-known to require any detailed treatment here. Varietal resistance or susceptibility to root disease caused by both obligate and facultative parasites is quite common. For example, varieties of potato resistant and susceptible to wart have been known for a long time. Invasion of all varieties does occur, but in those which are highly resistant, the pathogen fails to induce any symptoms and no resting spores or prosori are produced. In fact, a very wide

variability in response to infection occurs, some varieties producing no warts or only very tiny ones, and others inducing the formation of conspicuous and large warts. In the case of cabbage yellows due to *Fusarium oxysporum* f. *conglutinans*, two types of resistance are known, one of which (type A) is inherited qualitatively and the other (type B) quantitatively.

Susceptibility or resistance may often, but not always, be modified by the soil environment and by host nutrition. For instance, resistance of the tomato variety Marglobe to fusariose wilt is markedly influenced by the type and concentration of host nutrition, but the high resistance of the red-currant tomato is not affected. Similarly, with cabbage yellows, resistance of type A plants is not affected by host nutrition, whereas in both the type B resistant plants and susceptible plants disease percentage is inversely proportional to concentration of nutrients and to level of potassium supplied.

3. *The Soil: Physical, Chemical, and Biotic Factors; Cultural Practices*. That the environment has a marked influence on infection and pathogenicity is well known. In the case of root diseases, the soil environment is of particular importance. This environment includes not only the rhizosphere of the host plant but also the soil away from the rhizosphere.

The soil environment reflects the combined effects of a number of factors: physical factors such as soil type and texture, moisture, aeration, pH, and temperature; chemical factors which include the mineral status (nitrogen, phosphorus, potassium, and heavy metals) and organic matter of the soil; and biotic factors. Besides these, the soil environment is usually modified also by cultural practices. Recent research shows that the soil environment does not usually permit an easy analysis of its effects. In the case of fusariose wilt of pea, for instance, 28°C. is the optimum temperature for the growth of the pathogen, but the temperature-disease curve rises from a minimum at 15° to an optimum at 21° and then falls again to another minimum a little above 30°. In sand culture, however, the temperature-disease curve shows a peak at 28°, conforming largely to the temperature-pathogen curve. These observations are suggestive of the operation of the microbial factor as a modifying influence on pathogenicity in soils alone but not in sand culture. Similarly, the temperature-disease curve may also be modified by the nature of host. This effect is well seen in the case of *Gibberella zeae* causing seedling blight of wheat and of corn. The temperature-growth curve of this pathogen is similar to that of the pea wilt *Fusarium* and appears not to be directly related to the temperature-disease curve. This curve shows an optimum below 20°C or above for wheat, a low temperature plant, and an optimum below 20° for corn, a high temperature plant. Thus, with both hosts, the fungus is least pathogenic at the optimum temperature for the growth of the host.

The interaction between the various physical, chemical, and biotic factors of the soil in influencing root diseases is best illustrated by data obtained on soil microfloras and their effects on soil-borne diseases. Cotton wilt, due to *Fusarium vasinfectum*, has been intensively studied from this angle and a summary of the highlights of the results obtained will prove very instructive here. In India, wilt is confined to distinctly alkaline soils with a pH of 8–9. The addition of certain heavy metals such as boron, zinc, iron, and manganese is known to control wilt and there is evidence to show that the microbial numbers in the rhizosphere and in soil away from rhizosphere increase considerably in the presence of these amendments. It has further been shown that the addition of aluminum, lithium, boron, zinc, manganese, and iron to these soils also limits saprophytic survival of fusaria on plant debris. These fact indicate that reduction in pathogenicity which accompanies addition of these heavy metals to the soils is due at least in part to increased microbial antagonism to the pathogen in the soil and in the rhizosphere, although the possibility of increased host vigor and resistance cannot be ruled out. A point of further interest here is the controlling influence of pH and of combinations of these elements on their effectiveness in root disease control. With iron and manganese in 2 : 1 combination (40 p.p.m. iron : 20 p.p.m. manganese), for instance, maximum increase in microbial numbers in soils and greatest protection to cotton plants from wilt were both seen at pH 6. It is also known that microbial activity in these soils as well as saprophytic survival of *Fusarium vasinfectum* therein are interrelated and these, in turn, are modified by soil temperature, soil moisture, soil aeration, and also the nutrient status of the soil. That addition of organic nitrogen in the form of organic manure may also give protection to plants from wilt is evident from the work of McRae and Shaw on wilt of pigeon pea and of several workers on cotton wilt; even here the protection seen appears to be the result of increased microbial antagonism to the pathogens.

In the case of soil-borne tobacco necrosis virus, the observation that host plants growing in bacteriologically sterile conditions and in water culture are not infected, but contract the disease in soil that has been recently sterilized, may be interpreted to mean that some members of the soil microflora serve as hosts for the virus; however, Bawden (1950) records that no experimental evidence for this hypothesis could be obtained.

Finally, it may be mentioned that synergistic effects between pathogens are also common and in many cases root diseases may be caused by mixed infections by more than one pathogen.

4. *The Rhizosphere: Pathogen in the Rhizosphere and Its Relation to Other Rhizosphere Micro-flora, the Microflora of the Soil, Root Exudates, etc..* It is probably well to begin by defining rhizosphere. A soil ecologic region inside which the soil is subject to specific influence of plant roots is the rhizosphere; the credit for this definition goes to Hiltner (1904). It is a matter of surprise that the import of this impact of the growing root on the soil medium through which it grows failed to draw much attention until the thirties when Starkey (1929) revitalized the subject and did real creative thinking in this fascinating field of research. Since then, much work has been done and many excellent reviews on the subject have appeared. Consideration of the rhizosphere as a region of intense microbial activity would logically concede that the root surfaces are normally a more potent source of energy than the soil adjacent to plant roots and to designate this the term "rhizoplane" has been suggested which includes in its definition external surfaces of plant roots together with closely adhering particles of soil or debris. Other terms such as "outer rhizosphere" and "closer rhizosphere" have been used to designate sites of microbial concentration. Despite these newer terminologies, more recent investigations into many aspects of this complex problem of root exudates and microbial activity seem to favor and justify the use of just two terms, the rhizosphere to denote soil region adjacent to plant roots and the rhizoplane to indicate plant root surfaces. There is, however, need for familiarity with one more technical term, the "rhizosphere effect", since on a proper understanding of it depends the quantitative assay of rhizosphere problems. The rhizosphere effect is the ratio of the number of organisms in the rhizosphere (which in our view includes the rhizoplane) and the number in the soil outside the rhizosphere, calculated on a soil dry weight basis. This is generally expressed as a positive effect if the ratio exceeds one, and negative if fractional.

Rhizosphere or rhizoplane microfloras and their relationship to production of antibiotics *in situ* have naturally broadened the field of inquiry and many new techniques have been evolved for detecting these products of microbial synthesis but it would appear futile to cover that ground, since our immediate task is to understand the rhizosphere in relation to root disease pathogens in as broad a sense of the term as possible. The building up of an active rhizosphere or rhizoplane in germinating seedlings does not take long and, indeed, Timonin (1940) showed that seedlings only 3 days old had 11 to 28 times as great a rhizosphere population as elsewhere in the same soil. It is now known that microorganisms seldom occur on the root tip and their appearance is governed by normal plant development rather than root growth. In other words, it is intimately connected with active and normal plant metabolism and, therefore, primarily on root exudations. Evidence indicates that this process is not a one way affair, in fact, the activities of microorganisms in the soil of the rhizosphere include decomposing of sloughed-off root caps, root hairs, cortical and epidermal cells, and making available organic and inorganic nutrients for absorption. The soil inhabitant class of organisms and the soil invaders have, therefore, an important part to play in these processes and together these floras constitute what one might designate as microecology in the broadest sense of the term reflecting as it does on the products of metabolic functions of the rhizoplane and the rhizosphere. Preliminary observations on the properties of powerful antibiotics derived from the plants *Aucuba japonica* and *Myrtus communis* indicate their potential influence on the ecology and sociology of the microbial population of the rhizosphere.

Working with tropical soils, Agnihothrudu (1954) showed that the rhizosphere of the wilt-susceptible variety of pigeon pea, *Cajanus cajan* (Spreg.) Millsp. was more conducive to the survival of its causal agent. *Fusarium udum*, in contrast to the resistant varieties. Further, the actinomycetes antagonistic to *F. udum* were present in large numbers in the rhizosphere of the resistant varieties. The significance of this finding is becoming more obvious with recent investigations comparing the rhizosphere floras of resistant and susceptible varieties of crop plants.

It was shown earlier by Agnihothrudu (1953) that *Aspergillus* spp. predominated in the rhizospheres of potted plants, particularly *Sorghum dochna* var. *irungu* and cotton, from 15 days to 3 months after germination. In general, both in the unplanted soil and in the rhizosphere the percentage of *Aspergillus* spp. decreased toward the final estimation except that there was an increase in those plants that flowered, viz., *Phaseolus vulgaris, Cyamopsis tetragonoloba, Sesamum indicum*, and *Crotalaria juncea*. *Penicillium* spp. also increased gradually in numbers as the plants grew older. The changes in total numbers of fungi corresponded roughly to the changes in numbers and percentage of *Aspergillus* and *Penicillium*. *Fusarium* spp. were encountered regularly in the rhizosphere dilutions but very rarely in the control soil, the highest increase occurring with pigeon pea and the lowest with *S. dochna* var. *irungu*, possibly indicating an inhibitory root exudate in the case of sorghum. *Macrophomina phaseoli* and *Neocosmospora vasinfecta* were observed microscopically in the rhizoplane of all seedlings except sorghum but the organisms did not grow in dilution plates. In addition, eleven other genera of fungi were obtained from the rhizosphere, and in plants that flowered, there was a definite increase in the numbers and percentage present at the final estimation.

These results have found support in a more recent investigation into the rhizosphere floras of pea infected by *Fusarium oxysporum* f. *pisi*. Four pea cultivars which are differential hosts for the physiologic races of the pea wilt fungus *F. oxysporum* f. *pisi* exerted differing effects on the soil microflora. The cultivar susceptible to race 1 supported more fungi, bacteria, and actinomycetes near its root surface than do the cultivars that resist race 1. Spores of a race that can cause wilt to a particular cultivar germinate well in oil extract from that cultivar, whereas germination decreases in extracts from rhizospheres or in root exudates of a resistant cultivar. Similarly, the susceptible cultivar wilted severely where the susceptible cultivar was previously grown and inoculated with race 1, whereas wilting was less and developed more slowly when the cultivars resistant to race 1 had been previously cropped. Possibly, substances exuded by roots of cultivars resistant to race 1, prevent race 1 from germinating.

Sulochana (1958) has recently undertaken a quantitative study of the microfloras of the rhizosphere of cotton plants and has used reliable bioassays for analyzing exudates. The results are of far-reaching significance. The choice of soil was confined to the heavily wilt-sick soils from southern India where as many as fourteen species of pathogenic fusaria have been isolated before, the most predominant species being *Fusarium vasinfectum*. The rhizosphere floras of two genetic strains of two species of cotton (the diploid susceptible *Gossypium arboreum* race *indicum* L. and the tetraploid resistant *G. hirsutum* L.) were examined. In both species the amino acid requiring bacteria were qualitatively larger in numbers than the vitamin requiring bacteria. Quantitatively, higher numbers of both groups of bacteria existed in the rhizosphere of diploid strains than that of the tetraploid strains. The bacterial numbers of both groups diminished as a result of the rhizosphere effects of wilting susceptible diploid plants in soil when the pathogen was present. This indicates that much of the energy substances exuded had been utilized by the pathogen *F. vasinfectum*. Quantitative assays for amino acid were made using the test organisms *Lactobacillus arabinosus* 17/5 and *Leuconostoc mesenteroides* P 60. Concentrations of amino acids were higher in the exudates of diploid strains than the tetraploid strains. Assays of the vitamin B group using X-ray mutants of *Neurospora crassa* and *N. sitophila* indicated that vitamin B concentrations were considerably higher in the vicinity of the diploid than of the tetraploid plants. In general, exudation of amino acids and vitamins could be directly correlated with the rhizosphere activity. On growing susceptible diploid variety of cotton in garden soil at constant temperatures of 32.5°, 35.0° and 37.5°C., no appreciable effect in the rhizosphere of healthy and infected plants was observed at the two higher temperatures. This is not surprising. At these high temperatures an abnormal host metabolism could result in a minimum of exudation of energy substances utilizable by the rhizosphere microorganisms. This could be the deciding factor in the observed marked reduction on both the amino acid and the vitamin requiring groups of bacteria. The energy material exuding from the root systems of genetically differing Old and New World cottons may well form the springboard for the development of saprophytic antagonistic bacterial floras and of mycofloras as well. These form part of the rhizosphere flora and are known to act as root-infecting pathogens. We have not examined sufficiently the implications of gene controlled mechanisms in light of how they affect uptake and exudation of metabolites. Quite obviously, the plasma membranes of root systems are worth studying more critically since in them rests the key to exudates and their repercussions on saprophytic and parasitic microfloras of the rhizosphere and, more particularly, the rhizoplane. Apart from exudations of amino acids and sugars, substances of large and small molecular weights such as nucleotides and flavanones have been recorded from pea exudates.

Studies on the nature of root exudates of many crop plants under varying soil conditions have been undertaken by numerous workers. Katznelson et. al. (1954) showed that desiccation and subsequent rewetting of the sand in which tomatoes, soyabean, barley, or oats grew, resulted in the excretion of glutamic acid, aspartic acid, leucine, alanine, cysteine, glycine, lysine, phenylalanine, proline, and a reducing compound of R_1 value identical with glucose. Rovira (1956a) indicated that, under aseptic growing conditions in quartz sand, peas excreted twenty-two different amino acids while oats excreted fourteen of them. He further indicated that addition of exudates from the roots of peas and oats *in vitro* increases the growth of microorganisms isolated both from control soil and the rhizosphere of 3-week old pea plants. This stimulation was greater for organisms from the rhizosphere than for those from outside the zone. Some organisms responded in a comparable way to addition of yeast extract, whereas others did so with exudate plus yeast extract. The root exudate was not totally replaceable by glucose, soil extract, vitamine free casamino acids, or a synthetic mixture of growth factors. Stimulating action of yeast extract on microorganisms is generally considered to be due to unidentified factors.

Bhuvaneswari and Subba-Rao (1957) examined the root exudates of *Sorghum vulgare* var. *dochna* and *Brassica juncea* and spotted several organic acids and sugars in them. Although tartaric and oxalic acids, D-xylose and D-fructose were common to both plants, *B. juncea* had malic and citric acids, and D-glucose and maltose in addition. The possibility of malic and citric acids as factors influencing the observed depression of the microflora of *B. juncea* has been suggested by these authors. Root exudates of paddy infected by the foot rot organism, *Fusarium moniliforme*, have been the subject of study recently. Chromatographic analysis of exudates from susceptible and resistant strains of paddy under inoculation revealed that aspartic acid, gultamic acid, tryptophan, and lysine were common to both plants and occurred in almost the same concentrations. In addition, the resistant strain had cystine, asparagine, tyrosine, and methionine. Especially the sulfur containing amino acid cystine has been shown in this laboratory to be present in fairly large quantities in the roots of the tetraploid variety of cotton, *Gossypium hirsutum* L., which is resistant to wilt caused by *Fusarium vasinfectum*. Studies of the rhizosphere microfloras of these two strains of paddy showing differing amino acid exudation would be of considerable interest.

An interesting example of possible root exudation and its effect on pathogenic fungi comes from the work of Ellis (1951) who demonstrated that the clubroot fungus (*Plasmodiophora brassicae*) was greatly reduced or was eliminated when susceptible cabbages were grown after peppermint (*Mentha piperita*) had been cropped from 1 to 3 years. The author concludes that the stolon of *M. piperita* was perhaps indirectly responsible for the production of substances antagonistic to *P. brassicae*. On the other hand, there may be direct correlation between exudation from *M. piperita* and *P. brassicae* antagonism. This problem can, therefore, be pursued further. Root exudates of *M. piperita* could be examined for their ability to cause widespread soil disinfection of the otherwise persistent pathogen.

There are also other functional advantages that rhizosphere microfloras can confer in the normal uptake and utilization of inorganic salts by roots from soils. Gerresten (1948) reported that roots of oats, mustard, sunflower, and rape with a rhizosphere population were capable of absorbing and utilizing insoluble mineral phosphates which were only slightly available to sterile roots. There is also evidence from other sources that, by absorbing the water-soluble phosphorus compounds, the bacteria prevent to a certain extent the chemical decomposition of these compounds in soil. Varieties of oats susceptible to manganese deficiency supported greater numbers of microorganisms that showed capacity for oxidizing manganese into unavailable forms than the resistant varieties. Conversely, microorganisms have been found to reduce availability of microelements when energy materials are added to soils. Eliminating the natural microflora by sterilization of soils cures certain microelement deficiency symptoms, the effect being a direct consequence of removing the microbes that act as competitors for the meager supply of micronutrients essential for normal plant growth.

There may be many other functional mechanisms of ionic uptake by higher plants in both acidic and alkaline soils, not only of macro but of microelements also, where the role of microorganisms has to be determined by careful experimentation before a verdict is given. In this, much ingenuity in evolving techniques has to be displayed, particularly in the collection of root exudates under aseptic conditions. Bioassays must be more widely developed for these investigations. Without them, the dynamics of the rhizosphere and rhizoplane could hardly be expected to progress beyond a stage of stalemate.

D. Root Infections and Symbiosis: Mycorrhiza, Root Nodule Bacteria

We cannot accept that all infections of roots that are beneficial to their hosts are mycorrhizal or that the term "mycorrhiza" specifically pertains to proved examples of symbiosis. Both aseptate and septate mycelial forms of great diversity enter into this symbiotic relationship with rhizoids and root of bryophytes, pteridophytes, and angiosperms. Collectively, these constitute the endotrophic mycorrhizas as opposed to the ectotrophic that have mycelium external to the roots. The ectotrophic form has a well developed sheath of fungal pseudoparenchyma enclosing the root, the fungal hyphae penetrating between cortical cells, but few entering them. By contrast, in the endotrophic form the fungus is in the tissues, particularly in the cortex where the host protoplasm digests it. The aseptate fungi produce both arbuscules and vesicles that aid the digestion of the host tissues. The type of digestion varies from thamniscophagy (arbuscule and sporangiole digestion) to tolypophagy (formation of digestion clumps). The terms mycorrhiza and pseudomycorrhiza are at present not precisely definable. Infected roots showing ectotrophic mycorrhizas, as in the case of forest trees, exhibit constant morphological form, i.e., presence of external mantle and internal Hartig net (the central core of host tissue which is penetrated by the fungal hyphae in its outer cortex growing among the cells to form a network is the Hartig net, named after the German botanist), hypetrophy of the cortex, and characteristic branching. The pseudomycorrhiza covers a wide variety of endotrophic and ectotrophic associations which depart in morphological form or physiological function from the typical cases. In other words, the pseudomycorrhizal as well as the mycorrhizal ectotrophic forms show a microorganismal population in the rhizoplane and the rhizosphere in which one or a few fungi dominate in functional activity. For more detailed treatment of this and related literature, the reader could most usefully study the splendid reviews on the subject by Harley (1952).

The occurrence of digestion *in vivo* in the typical endotrophic forms no doubt demonstrates an exchange of material between the host and the fungus but its absence does not prove to the contrary, indeed, there are other modes of exchange that need not be preceded by digestion. The degree of dependence of the fungus on the soil for its vital metabolic needs in the shape of organic and inorganic nutrilites has not been precisely defined. However, it is generally understood that the penetration and infection of young roots by the fungus is from the soil or from plant residues or by root contact.

Many members of the Agaricaceae, Boletaceae, *Scleroderma*, etc., form mycorrhizas. Recent investigations by Robertson (1954) have shown that some of these produce air-borne basidiospores, e.g., *Boletus granulatus*.

Apart from the theoretical aspects of mycorrhizal infection and their importance in plant growth, the problem is one of great practical application, when fully understood. In the presence of appropriate fungi, under artificial inoculation conditions, there is undoubtedly great increase in growth of seedlings. The probable mechanisms involved in this increase are: that the fungi may alter the insoluble inorganic carbon compounds into soluble form, or change the pH in the culture medium, or produce vitamin-like substances (auxins?). This faculty is not confined to mycorrhizal fungi, however.

There is yet no proof that infection of the root tissue and the growth of the endophyte into the tissue are essential for host stimulation and many believe that infection is purely incidental. (The dependence of seeds on biological stimulation, whatever its mechanism, does not appear to offer an explanation for the formation of mycorrhiza in later stages). Nevertheless, it is difficult to discountenance the need for an external food base wherefrom the fungus would derive sufficient energy to enable it to develop into an active endophyte. The ability of endophytes to grow *in vitro* does not in any way indicate that there exists a competition for organic matter in soil. Burges (1936) states that infection is unimportant and the external activity of fungi is all important in the nutrition of mycorrhizal hosts. It would then appear that the emphasis almost completely shifts to the life of the endophyte in the soil, particularly in the rhizosphere and the rhizoplane. The problem of the endotrophic mycorrhiza, therefore, appears to be one of great complexity, where the food base in the soil from which the fungus meets its energy requirements and the conditions under which it parts with it to host tissue *in vivo* are logical steps in experimentally elucidating the sequence in the process. The main point to appreciate in the case of the ectotrophic mycorrhizas, on the other hand, is that a high proportion of absorbed materials must necessarily enter the infected roots routed through the fungal sheath as indeed, the fungal sheath is a living entity with powers of selective absorption. The completeness of the living fungal sheath and its intimate connection with the root cortex in mycorrhizal associations are of paramount importance in planning further experimentation to understand this functional physiology.

Apart from these general considerations, a number of fundamental investigations on the physiology of the mycorrhiza needs discussion in the present context. Harley and McCready (1952) examined excised mycorrhizal root tips of beech, separating the fungal sheath from the core after absorbing radioactive phosphorus (P^{32}) from aerated media at pH 5.5. Approximately 90% of the phosphorus accumulated in the fungal sheath. No significant change in the ratio occurred over a 24-hour period using a concentration of 1 mg. per liter. Cores freed from their sheaths before exposure to phosphate absorbed phosphorus four times more quickly than the cores of intact tips. Harley and Brierbley (1954) further showed that the route by which P^{32} passes through the fungal tissue into the host tissue of beech mycorrhiza was from the sheath into the core. The rate of movement of P^{32} from fungus to host was rapid at first but became slower after 15 hours and the rate was temperature sensitive, becoming very slow at 1°C. The process of transport was sensitive to oxygen concentration, slackening very much when oxygen concentration in solution was below 3%. It was obvious that active transport of phosphorus from fungus to host occurs in mycorrhizal beech roots and the mechanism of transport is dependent on aerobic metabolic processes in the fungal tissue as well as on the absorptive processes of the core. Phosphate lost from roots kept in low oxygen concentrations is entirely released from the sheath and this results from temperature sensitive anaerobic processes occurring in that tissue.

Studying phosphate uptake further, using P^{32}, it was shown that excised mycorrhizal roots of the beech when washed in buffer solution containing phosphate recorded reduced rate of transport from fungus to host. When roots had been returned to phosphate free buffer, after a period in buffer containing phosphate, the temporarily reduced active transport was resumed at a rapid rate. It is suggested that the phosphate absorbed from the external solution competes successfully for a substance produced in respiratory metabolism so that phosphate accumulated in the sheath remained immobilized. In the absence of external phosphate supply, the phosphate already accumulated in the sheath was utilized. Movement of phosphorus into the core was mainly from the external solution when this contained phosphate, but was from the sheath phosphate when the external supply failed. Earlier, Harley and McCready (1952) had shown that the fungal sheath constituted about 39% of the dry weight of mycorrhizal roots of beech. The sheath of intact mycorrhizas restricted the uptake of phosphorus by the cores—and this reduction was less at higher than at lower concentrations—acting as a partial barrier to diffusion and phosphorus absorption seemed to be linked with metabolic activity over the whole range of phosphorus concentrations studied There are two possible routes by which phosphorus may reach the core by diffusion through the intercellular spaces and cell walls of the sheath and by way of the living sheath cells.

That the physiological state of metabolism of the plant has much to do with mycorrhizal function finds support from the work of Harley and Waid (1955). The influence of different levels of daylight radiation on the growth and nature of mycorrhizal infection of beech seedlings was direct, the mycorrhizal infections on the root systems appearing after the first true leaves emerged. Plants receiving more light showed more vigorous development than shaded plants. Shading eventually led to a loss of resistance by the beech seedlings to parasitic infection while increases in light intensity resulted in mycorrhizal formation.

There is yet another aspect of nutritional requirement by the mycorrhiza-forming fungi. Melin (1954) tested a number of Hymenomycetes and Gasteromycetes and all of them were partially or totally deficient in thiamine. Most species were heterotrophic for both the thiazole and pyrimidine fractions while five species of *Tricholoma* were more deficient in pyrimidine than in thiazole. Some species of *Cortinarius* had a reduced capacity for synthesizing thiamine from its two components, thiazole and pyrimidine. Most Basidiomycetes greatly benefited by the presence of small amounts of amino acids in ammonium nitrogen containing medium. The main sources of these nutritional substances in nature seem to be the soil and not the roots. *Boletus* spp. grow extremely well *in vitro* on excised *Pinus sylvestric* roots in a nutrient solution supplemented by amino acids plus the B vitamins, whereas without roots growth was barely visible. Pine roots seem to produce one or more growth-promoting metabolites essential for the growth of tree mycorrhizal fungi which are deficient in these substances. Excised tomato roots and germinating pine seeds affect the growth of these fungi in the same way as pine roots, indicating that the metabolite is not specific to pine roots.

Much has been said to indicate the somewhat parallel nature of the problems of the saprophytic and parasitic forms of the rhizosphere and the mycorrhizal habit as far as the interdependence of host metabolism is concerned for both types of infections. Quite obviously, much of the energy substances exuding from roots are utilizable by the micropopulation and it is probably premature to draw a line and compartmentally assign these microbial functional processes. There seems little doubt, however, that this twin field of research would give endless opportunities for more critical work on absorption and movement of metabolites from the soil into the plant tissues and back as exudates and their utilization by facile endotrophic and ectotrophic living microbial populations.

The root nodules of Leguminosae and the bacteria that partake in this symbiotic relationship present another interesting facet of soil microbiology. As early as 1587, Dalechamps named a species *Ornithopodium tuberosum* which could be distinguished by its nodule-bearing habit. The practical significance of these nodules was first noted by Hellriegel in

1886 when he mentioned that nitrogen fixation in the leguminous plant occurred only when root nodules were present. Since then the leguminous nodules have been studied intensively by many workers and it will be pertinent subject.

The general assumption that nodulation ability is characteristic of all Leguminosae is probably not correct. Indeed, only less than one-half of the genera in this order have been examined and of these many species falling under Caesalpinioidease do not possess the ability to form nodules. The greatest number of species having this ability belong to the subfamily Papilionatae. Compatibility of the bacterial and plant proteins, and a direct relationship with the calcium fraction of the leguminous plants (this fraction being greater in leguminous than in nonleguminous plants) have been suggested to explain why rhizobia produce nodules only on leguminous plants. It has also been pointed out that the leguminous plant produces an enzyme which enables it to select, entrap, and use the particular form of organic nitrogen contained in the invading rhizobia. Several workers have also recently demonstrated resistance to nodulation by pure lines of genetically selected red clover and soya bean which indicates that a recessive hereditary factor in the plant may also be involved.

The rhizobia are facultative parasites but their role in symbiotic nitrogen fixation has led to their being regarded as symbionts. Culturally, all strains are aerobic, heterotrophic, gram-negative rods. Although soil is the normal habitat of these bacteria, it is extremely difficult to distinguish them from other soil bacteria such as *Radiobacter* by ordinary methods. They are normal members of the soil microflora and can exist for many years in field soil without their host plant. However, its presence brings about a definite increase in the number of rhizobia in the neighborhood and this is possibly due to stimulatory root secretions. Strains differ in the rates at which they multiply around roots under the influence of these secretions. It is difficult to estimate the number of nodule bacteria in a natural soil, since the only definite test for *Rhizobium* is its ability to produce nodules on its host plant. However, approximate estimates of numbers obtained by supplying the plants with serial dilutions of the soil sample and finding out the extent of infection, were extremely variable, but these indicated that the numbers, for instance, of clover *Rhizobium* may be of the order of tens of thousands per gram in a soil of pH suitable for clover. Diversity of strains in a given soil is common. Proof is wanting as to the ability of rhizobia to fix atmospheric nitrogen in the absence of the host plant, in the soil, and in pure and mixed cultures *in vitro*.

In the understanding of the physiological relationships between the rhizobia and the host plants, two terms require definition: (1) infectiveness and (2) effectiveness. The former connotes ability of a strain to cause the formation of nodules on certain leguminous species and not on others; the latter, the ability of a strain to help the growth of the plant by nitrogen fixation. In each of these, various gradations in relationship also occur. *Effective nodules* fix quantities of nitrogen normally adequate for the plant's needs, whereas *inffective nodules* fix little or not nitrogen. These two types of nodules are known to show marked differences in the course of their development. In both types, early stages of growth are similar, the nodule first consisting of a tiny mass of meristem cells usually derived from the root cortex. Most of the central cells become infected and cease to divide. These cells are surrounded by a layer of uninfected cells which remain meristematic and form a distal cap. Owing to the activity of these meristematic cells, the nodule grows and the inner layers of newly formed cells are successfully invaded by bacteria. Vascular strands are later formed connecting the nodule with the central cylinder of the root; a secondary endodermis also appears. From about this stage, the two types of nodules differ. In the case of effective nodules, further growth takes place leading to the formation of a conspicuous mass of central bacterial tissue; in the case of ineffective nodules both the meristem and the central bacterial tissue are transient. In the latter case, the bacteria become parasitic on the host tissue and cause its disintegration and, indeed, this necrotic process extends distally through the central tissues until the middle of the nodule is destroyed and it ceases to function. In the case of the effective nodule, the central bacterial tissues is the seat of the symbiotic nitrogen fixation process. Further, it contains four well-defined pigments: leghemoglobin (red), legcholeglobin (green), legmethemoglobin (brown) and coproporphyrin (brown). Leghmoglobin is markedly similar to hemoglobin of blood and a molecular weight of about 34,000 (i.e., about half that of blood hemoglobin) has been reported for a preparation which approximated 85% purity; it was first identified by Kubo in 1939 and later confirmed by Keilin and Wang in 1945. Leghemoglobin does not occur in detectable amounts in ineffective nodules.

The function of hemoglobin and its derivatives in nodules remains an unsolved and intriguing problem, although there is circumstantial evidence connecting hemoglobin with nitrogen fixation. Neither the plants nor the rhizobia produce the hemoglobin independently, and it occurs only during the stage of effective symbiosis. It would appear that it is a product of the *rhizobium*-leguminous plant complex, one agent presumably contributing the hemin fraction and the other the protein fraction of the molecule. Its presence in effective nodules is closely linked with the photosynthetic activity of the host plant. In mature effective nodules it first appears during the differentiation of the bacterial tissue and can be detected histochemically only in rhizobia-packed cells, proof is wanting for its occurrence within the rhizobial cell per se, in the uninfected cells of the bacterial tissue, in the nodule cortex or within the meristematic area. Further, free-living rhizobia are not able to fix nitrogen following addition of hemoglobin and purified leghemoglobin. A further noteworthy fact is that no hemoglobin is produced in nonsymbiotic nitrogen-fixing systems. Space does not permit a detailed discussion of the biochemical aspects of symbiotic nitrogen fixation for which the reader should refer to the excellent reviews of Thornton (1954) and of Allen and Allen (1954).

Nutman (1946) has proposed the terms "responsive" and "unresponsive" for plant reactions to nodulation determined by genetic factors. However, it is far from clear whether the differences in plant response are due to the inherent qualities of the bacterial strains, to conditions within nodules, or to the influence of the host itself.

Infection of the leguminous root usually takes place through root hairs, although entry may be effected through epidermal and cortical cells and also through ruptured tissue at the points of lateral rootlet emergence. In the case of *Neptunia oleracea*, an aquatic leguminous plant lacking root hairs, epidermal cells appear to be the sole points of entry. Normally, a colony forms near the tip of a root hair; the latter excretes a substance (β-indolylacétic acid?) which causes the root hair to curve and at the bent tip the bacteria make way through the cell walls into the roo' hair. They then pass into the cells of the root, multiply rapidly and form the nodule. The number of nodules produced is characteristic for each bacterial strain, different strains exhibiting differences in infectivity. With a given strain and host plant, there is, nevertheless, correlation between the dose of bacteria supplied to the root region, the number of root hairs infected, and the number of nodules produced. Normally, host root systems may be simultaneously invaded by several strains of *Rhizobium* and different strains can be isolated from nodules on the same plant in the field. In the presence of more than one strain, the proportion of nodules produced by each strain will depend on (*a*) the relative numbers of each strain in the root region and (*b*) the relative infectivity of the strains. Nodulation produced by one strain may saturate the nodule-forming capacity of the host and when this happens a second strain may be excluded, although it cannot be concluded that inasion by one strain confers immunity to the host from subsequent invvasion by a different strain or strains.

Much more remains to be known about the mechanism of rhizobial entry into the root. Earlier suggestions about the secretion of the enzyme cytase by the rhizobia and consequent dissolution of the root hair wall lack experimental proof. It would appear that the curling of the hair is due to some secretion produced by the rhizobia before they enter. Thornton (1929) associated infection with production of a stimulatory substance by the roots at the time the first leaf unfold. During the period between germination and opening of the first true leaf of alfalfa, no infection occurred, whereas one day later 2% of the hairs were infected. Addition of the sterile extracts of solutions surrounding the roots of seedlings bearing first leaves to younger seedlings enhanced root hair infections. Production of the stimulatory substance is probably controlled or influence by the top of the plant, although removal of the first leaves does not apparently delay nodulation. Moreover, nodules may be formed on excised roots.

Nutman (1952) reported that with either lucerne or clover the actual numbers of nodules are greater on plants growing singly than on plants growing in pairs or in large groups within a culture of standard size. This inhibition of nodulation on plants growing together appears to be due to the diffusion of some substance from the roots, since the extent of the inhibition depends on the volume of the medium as well as the number of seedlings present. The number of infections per plant is directly proportional to the volume of the medium and inversely proportional to the density of planting. Turner's (1955) more recent work on nodulation in clover plants further supports Nutman's findings. It has been shown that the addition of charcoal to the rooting medium of clover plants inoculated with effective and ineffective strains of *Rhizobium* leads to a stimulation in nodule formation. In the presence of charcoal, the period between inoculation and the first appearance of nodules is reduced. One possible explanation of this phenomenon is that the stimulation is due to the adsorption by the charcoal of inhibitory compounds secreted by clover roots. These compounds have been eluted from charcoal and have been shown to influence nodule formation. Indeed, Virtanen and Laine (1939) have shown that aspartic acid, β-alanine, oxime N, and fumaric acid may be secreted from the nodule bearing roots of clover plants which are actively fixing nitrogen. Glutamic acid has also been detected. The influence of these secretions on nodulation is not known. However, repeated cultivation of clover crops in the same soil may lead to "clover sickness" which may be due to the accumulation of toxic compounds in the soil secreted by the clover plants; such sickness can be overcome by addition of charcoal to soil.

Apart from root excretions influencing nodulation, certain other factors may also have similar effects on nodule formation. For instance, invasion is often governed by the carbohydrate-nitrogen balance in the root hair. Thus, seedlings of etiolated vetch and albino soybean and *Leucaena* are not able to produce nodules unless an adequate amount of carbohydrate is supplied. According to Nutman (1946), resistance to infection of red clover by *Rhizobium* is attributable to a hereditary factor which delays nodulation as late as the unfolding of the fifth or sixth leaf. Results of experiments involving grafting of genetically susceptible clover tops onto roots of resistant plants and vice versa suggested no translocation of substances influencing infection.

The relation of the general soil microflora to nodulation has been well brought out by the work of Harris (1953) who investigated the plant-rhizobial relationships between *Trifolium subterraneum* and strains of *Rhizobium trifolii* in pot experiments. One strain (No. 44) was found to give an ineffective reaction (as measured by plant growth) in heat sterilized soils, whereas in similar nontreated soils an effective reaction was seen. In order to study the effect of associated fungi on nodulation, sterilized subterranean clover embryos were grown asceptically on a mineral salts soft agar medium free from nitrogen and carbohydrates. When the first trifoliate leaf had unfolded, the seedlings were inoculated simultaneously with *R. trifolii* and a suspension of one of the fungi frequently isolated from clover roots in nonsterile soils. The presence of a pathogen capable of killing the plant rapidly inhibited nodulation completely, as in the case of *Fusarium* 1013 and *Sclerotinia* 1015. Where the degree of pathogenicity was less marked but root invasion occurred in the extensive local root rots, as in the case of the dematiaceous fungus 1017 and *Hormodendrum N*, nodulation with strain 430 of *Rhizobium trifolli* was reduced, probably owing to impaired metabolism in the plant rather than direct antagonism of the fungus to rhizobia. Some strains of fungi and bacteria stimulated the weakly virulent strain 44, but had no effect on the more active strain of the bacterium.

These results point to the need for more comprehensive and intensive studies on the same lines and experimental work aimed at understanding the rhizosphere effect, vis-a-vis root exudates in nodule-forming leguminous plants would, indeed, be a most fruitous path for future studies. There is no doubt that the leguminous root nodule still poses many interesting and intriguing problems for the experimental microbiologist.

III. PATHOGEN-HOST RELATIONSHIP

The essential features of the pathogen-host relationships seen among root-infecting pathogens may now be considered. If one is guided by the damage caused to root systems or to the plant itself via root systems by organisms, it would appear erroneous to leave out of consideration the large number of organisms which are known to produce antibiotics, toxins, and other metabolites in soils which might predispose root systems to infection. However, it is necessary to maintain a distinction between such damage brought about by metabolic products of microbes and damage following infection by a pathogen. In what follows, therefore, the discussion will be centered on the pathogen in its relation to the host.

It is probably difficult to state which is the predominant factor in this dual relationship, the pathogen or the host. Nevertheless, it is safe to assume that the host, its rhizoplane, and its rhizosphere merely provide the substrate for the activity of the pathogen and, in the absence of the latter, there is neither infection nor pathogenesis. Pathogenesis, of course, implies that a pathogen must effect entry into the host and assimilate the available nutrients, must tolerate or overcome host resistance, and induce disease in the host by its action on host tissues.

What appears to be a simple type of relationship is that seen among some pathogens of the destructive type which kill the host tissues and derive nourishment from the tissues so killed. This behavior is characteristic of many facultative parasites and the whole course of events leading to pathogenesis would appear to rest on the ability of these pathogens to secrete enzymes capable of dissolving the host substrate and killing the cells. Enzyme studies on several root rot, soft rot, and damping-off pathogens have shown, for instance, that the three pectolytic enzymes, polygalacturonase, depolymerase, and pectin methyl esterase are invariably produced by several of these organisms. It is reasonable to expect in such cases correlation of pathogenicity with presence and amount of the enzymes produced. No doubt, such a correlation has been reported in some case, but not always. In studies on various species of *Rhizopus*, which are wound pathogens causing rot in sweet potatoes, for example, some of the pathogenic species (*R. nigricans* and *R. artocarpi*) secrete less pectinase than the two nonpathogenic species (*R. chinensis* and *R. microsporus*). Further, one species which produces the largest amount of the enzyme is not pathogenic to sweet potato. Similarly, it is difficult to explain why *Botrytis cinerea*, which rarely parasitizes potato, should be able to produce active enzymes in potato decoctions of various strengths, whereas *Pythium debaryanum*, which is pathogenic on ordinary mature potato tubers, shows negligible amount of enzyme secretion on ordinary potato decoctions. It is obvious that there are differences between the pectinase enzymes of *Botrytis* and of *Pythium* and, as suggested by Brown (1948), "the interpretation of these difference may be that the enzymes are different in themselves or that they are the same but that some of their properties are conditioned by other metabolites produced by the pathogenic fungi."

Apart from dissolution of cell walls, a feature characteristically seen in some of these primitive pathogens (e.g., *Botrytis cinerea* and *Sclerotinia scleroniorum*), is their ability to increase considerably the permeability to water of the host cells just beyond the discolored necrotic zone. Some substance other than the pectinase may bring about this change.

The pathogen-host relationship in common scab of potato due to *Steptomyces scabies* presents some noteworthy features. Following penetration through lenticels, stomates, wounds, or directly through the cuticle when it is thin, in young tubers of susceptible varieties many layers of dead cells are formed on the exterior in which the pathogen develops as a saprophyte. The underlying living cells then undergo abnormally rapid division and there is consequently sloughing off of more dead cells on which the pathogen continues to grow. It would appear that there is little invasion of living tissue and, indeed, the scab lesion results from proliferation of the latter. The mechanism of cell proliferation remains obscure. In resistant tubers, few dead cells occur at the surface and, therefore, the pathogen is not able to establish itself. Another interesting fact is that peel extracts from tubers examined chromatographically or by taking ultraviolet absorption curves showed a much higher concentration of chlorogenic acid in a resistant than in a susceptible variety. In some resistant varieties, the chlorogenic acid appeared to be more heavily concentrated in the periderm and particularly near the lenticels, which serve as natural avenues of entry for the pathogen, and also around mechanical or parasitic injuries.

Whitneny's studies (1954) on *Rhizoctonia crocorum* indicate that its ability to parasitize carrots requires substantial saprophytic development prior to infection. Its behavior, therefore, is somewhat comparable to that of *Streptomyces scabies*. The fungus infects mainly dicot hosts with roots which develop periderm tissue at some stage in their ontogeny. The susceptible stage of the carrot corresponds with that at which the carrot sheds its cortex in favor of a periderm. In such hosts, therefore, the pathogen can develop initially as a saprophyte on the periderm as a result of which it acquires the ability to launch a parasitic attack on the underlying tissues. In the invaded carrot tissue, intracellular sclerotial bodies are formed which appear to give rise ultimately to minute infection bodies. The intracellular hyphae are not haustorial and do not establish a nutritional balance within the invaded cell; on the other hand, the invaded cells are killed. In the organized host, conditions are not favorable for the more rapid fungus growth that usually occurs during its saprophytic development on agar and, therefore, under such conditions when nutrients are exhausted, the pathogen goes into a resting stage forming internal sclerotia.

The vascular wilt diseases involve quite a different type of relationship between pathogen and host. In the typical cases as, for instance, in fusariose wilts of cotton and tomato, following infection, the pathogen reaches the xylem and pervades this tissue and appears to cause little disintegration of the root systems and other living tissues. The complicity of enzymes, vivotoxins, and vessel plugging in pathogenesis of these plant wilts has been discussed earlier and by a number of authors and it would be superfluous to cover the ground again. Attention may, however, be called to the fact that among pathogenic forms of *Fusarium oxysporum*, most of which produce vascular wilts, some appear to produce cortical decay also. For instance, in watermelon wilt, at low temperatures, *F. oxysporum* f. *niveum* causes preemergence damping-off. This is a good example of the effect of the environment in modifying the pathogen-host relationship.

In the case of dry rot of potatoes caused by *F. avenaceum* and *F. coeruleum*, the latter grows through the intercellular spaces, the adjacent cells remaining alive, often for considerable periods—a situation somewhat suggestive of balanced parasitism. *F. avenaceum*, on the other hand, kills and penetrates the cells with which it comes in contact. This may be compared with what happens when blight-resistant potato tubers are inoculated with virulent and avirulent strains of *Phytophthora infestans*. Avirulet strains cause a rapid necrosis of the tissues with which they come in contact and are, thus, prevented from further spread or fructification; necrosis, however, is less rapid in tissues infected by virulent strains which are, therefore, able to spread through the tuber and fructify. These findings illustrate clearly the tendency exhibited by some pathogens to form a somewhat balanced association with the host. There is, of course, no question of mutual benefit here and it is a one-way traffic detrimental to the host. A greater approximation to balanced parasitism than what is seen in the case of *Fusarium coeruleum* is what characterizes the vascular wilt fusaria already referred to.

Some of the soil-borne smuts certainly exhibit a much better adaptation to live in a balanced association with their hosts. *Tilletia tritici* infecting wheat plants and *Ustilago violacea* infecting plants belonging to the Caryophyllaceae may be cited here. In both, seedling infection is common. In the case of *Tilletia tritici* germination of brandspores in soil is followed by penetration into the seedling. Following penetration, the hyphae grow intercellularly to the growing point and develop immediately behind it throughout the vegetative period. Later, the young leaves are invaded from the growing point, but only some mild symptoms appear at this stage. In the final stages, the flower primordia are invaded; the embryo is destroyed, the endosperm tissue is attacked, and spores of the pathogen are then produced in enormous numbers within the seed. The sequence of events in the case of *ustilago violacea* is similar, except that this pathogen ultimately affects the anthers and produces its spores within them. What is unique about both these diseases is the prolonged balanced association between pathogen and host from the seedling to the flowering stage.

The classical examples of true balanced parasitism are, of course, the rusts, and it is noteworthy that root infections by this group of pathogens appear to be very rare. Probably the only example of root infection by a nonsystemic rust is that recently reported in the case of *Puccinia carthami* which causes a foot and root disease of safflower.

We may now consider cases of root infection which result in marked hypertrophy or hyperplasia of host tissues. Clubroot of crucifers, potato wart, crown gall, hairy root, etc. are common examples. Crown gall, which is caused by *Bacterium tumefaciens*, has been studied intensively by a number of workers.

The mechanism of action of other root-infecting pathogens causing hyperplastic diseases, such as *Plasmodiophora*, has not been investigated in detail. In the case of potato wart, differences have been reported between warted and sound tubers; for example, the pH of the warted tissue is claimed to be higher than that of healthy tissue. Moreover, quantitative data on the ash content of healthy and warted tissue showed greater amounts of mineral constituents in diseased tissue, particularly of iron, manganese, copper, and nitrogen—an observation which probably indicates that the stimulus to hypertrophy results from the diversion of these substances to the seat of infection.

Mention may also be made here of some interesting observations on associations between plant roots and viruses. In some cases, such as the phony disease of peach, viruses appear to be localized within root systems without producing, however, any visible symptoms. On the other hand, in wound tumor disease, the roots have many spherical woody tumors which vary in size in different hosts; the largest tumors are formed in sweet clover plants (*Melilotus alba* and *M. officinalis*) and in *Rumex acetosa*. Wounding appears to be essential for the initiation of these tumors. Moreover, these tumors, have been shown to be capable of indefinite growth as tumor tissue. In the little known clubroot disease of tobacco, the distortion of smaller and larger roots, closely similar to clubroot of cabbage, is the main symptom. Death of lateral roolets is common in some tree diseases if virus origin, e.g., elm phloem necrosis. Death of lateral roots is also a common secondary symptom in tristeza disease of citrus, the roots being starved due to interruption of normal food transport. The disease occurs only when sweet orange scions are grafted on to sour orange stocks; the trees, however, hardly show symptoms characteristic of virus infections and they apparently die from a root rot. This rotting of roots appears to be a secondary effect resulting from the death of the phloem cells at the stock-scion junction and the consequent interruption in translocation of food from scion to stock. The sweet orange apparently is a carrier of the virus to which the sour orange is hypersensitive. When infected, the scion manifests no symptoms; movement of the virus into the stock, however, causes local reactions of a necrotic type which interfere with translocation of food, so that ultimately the whole tree dies.

From what has been stated it will be evident that the subject of pathogen-host relationship, as far as it relates to root-infecting pathogens, is a complex one, involving as it does interactions between the soil, the pathogen, and the host. No doubt, the facultative parasite has received more attention in studies of relationship between pathogen and host, but our knowledge of this relationship in the case of some of the obligate root-infecting pathogens is meager. There are, therefore, many possibilities of worth-while future study in this interesting group of plant disease organisms.

IV. PROSPECT

In what has been written so far, an attempt has been made to cover in brief outline some of the basic facts and speculations relating to the interactions of pathogen, soil, other microorganisms in the soil, and host. The rapid progress made in recent years toward a better understanding of these interactions has ushered in many new ideas such as the concept of rhizosphere and rhizoplane microfloras and of vivotoxins. There is no doubt that future work will have to be largely molded by these two concepts which need to be fortified further by critical study. A newer approach to investigations on the rhizosphere is called for here which should take into account not only root exudations, and the impact of microbial metabolic products formed *in situ* in soils on root-infecting pathogens and the rhizosphere, but also the ionic balance or imbalance and osmotic changes within living and infected plant roots. In the present state of our knowledge, or ignorance, it would appear that the ionic status of many of the key elements in the living healthy and in diseased plants would have to be determined accurately, since chelation between heavy metals and products of host-pathogen interaction has been postulated. We do not yet know if such chelation is an invariable feature within living plant systems, but if it really is, then possibilities of such chelation mechanisms leading to essential heavy metal starvation in plants would have to be seriously reckoned with. After all, vivotoxins production and its complicity in disease would depend on the host substrate and we do need detailed information on these. It is to be hoped that at least some of these problems would be taken up in the not distant future and, with the extraordinary refinements in modern biophysical and biochemical techniques such as spectrochemical and chromatographic analysis, radioactive tracers and tissue culture, and horizon of our knowledge can be widened considerably.

Similarly, root-infecting obligate parasites also call for intensive study, since our understanding of the physiology of host-parasite-soil interaction in this group of pathogens is appallingly meager. What, for instance, is the difference between healthy and warted potato tissue?

What is it that stimulates tumor formation in potato infected with wart disease, or tobacco plants suffering from virus clubroot? Or again, do normal healthy roots of cabbage differ in their exudates and rhizosphere microfloras from cabbage roots infected by *Plasmodiophora?* These are questions for which we have at present no answer. It is true that extensive studies have been carried out on the physiology of parasitism of leaf-infecting obligate parasites such as the powdery mildews and the rusts; but tne root-infecting pathogens have been left behind. Studies of rhizosphere microfloras, microclimatology, root exudates, and of *in vivo* changes in ion accumulation, growth substance levels, permeability, osmotic changes, cellular oxidase systems, etc., will, therefore, have to be extended also to these root diseases caused by obligate parasites.

There is another fundamental problem of great interest that invites urgent attention. It has been suggested time and again that plant viruses, particularly the so-called soil-borne viruses, may have some relationship with soil fungi or other soil microorganisms for their multiplication. For instance, tobacco necrosis virus has been shown to survive in soils, but just how it does so remains unsolved, although the possibility has been visualized of soil microorganisms serving as hosts. It appears that this problem will have to be worked upon in considerable detail if we are to arrive at a solution. The idea in bringing root-infecting viruses into this discussion is to enlarge our experimental outlook for the future. In the event of viruses being shown to have intermediate hosts among microorganisms, the question would then arise as to how, when other than the angiospermous hosts, the gymnosperms, pteridophytes, and bryophytes have not been known to suffer from virus diseases, the microscopic members of the plant kingdom could be infected.

It may be conceded, no doubt, that, by the nature of the very material which requires study, experimental investigations on these problems are beset with many difficulties. Nevertheless, with the aid of suitable modern techniques it should be possible to take up these problems with confidence and imagination and study them vigorously and intensively.

3

AUTONOMOUS DISPERSAL

For the purpose of this autonomous dispersal is interpreted as the spread of vegetable, fungal, bacterial, virus, or nematodal plant pathogens through the agency of soil, seeds, or plant parts in the normal practice of crop husbandry and in the distribution of plants and plant products, and not through the intervention of any extraneous agency such as insect, wind, or water. Such dispersal involves the consideration of both space and time; adaptation for dispersal and the consequences of adaptation are also discussed. The subject is dealt with under the three main headings of soil, seeds, and plant parts.

I. SOIL

A. General

Soil is the medium in which all the world's major crops are grown, and to the agronomist it means the substratum in which the crop plant is anchored by its root system and from which water, nitrogen, and mineral nutrients are derived. The basic structure of soil is a rock complex, be it coarse sand or fine clay; this part of soil is inanimate and inert; its function is almost completely physical, and it provides the anchorage for the plant's root system. Whether originally formed by the breakdown and crumbling of igneous rocks or by sedimentation, this inert medium, bathed with water and dilute salt solutions, forms the basis for the colonization of the earth by plant life. Long before the intervention of man, the algae, liverworts, mosses, ferns, and flowering plants gradually occupied and spread over this rock matrix to build up the wild flora of today. But with the death and decay of succeeding generations of plants and the attendant wild fauna dependent upon the flora for its existence, the dead, unyielding rock, sand, and clay began to develop a new look: and there became introduced into it a new, near-living component, which may be loosely described as its humus.

It is this humus content of the soil which is of such greater importance in agriculture and horticulture; not only does it improve the condition of the soil by ameliorating its physical structure, but it provides much of the food for ensuing generations. The dead bodies of plants and animals do not break down and decay to form humus as the result of slow inanimate oxidation or combustion. The death of one organism quickens the life of another, and so it is that the process of degeneration and decay is assisted and hastened by the advent of countless saprophytic, chlorophyll-less organisms, mainly microscopic, which depend for their very existence upon organic food made available through the death of previous generation. It is through the action of this multitude of microsaprophytes that ash again becomes ash and dust becomes dust. The introduction of humus to the soil converts it from an inanimate and inert matrix to a medium seething with life ecologically varied according to the origin from which it was derived. It is this living nature of the soil of farm and garden which is now attracting greater attention, and the microbiologist, finding here a field of work not only of intense academic interest but also one of great importance in crop husbandry, comes to the aid of physicist and chemist in the study of soil science. It is true that in hydroponics, where the plant is grown in a bath of sterile nutrient solution in pure liquid form with or without a gravel base, the problem of humus does not arise, but until such practice becomes much more widespread, the soil as such must remain the husbandman's main medium for crop growth. That the bulk of the micro-organisms which go to make up the flora and fauna of the soil are, on the whole, beneficial and play an essential part in building up soil fertility is abundantly clear—e.g., the action of the nitrifying bacteria—but the soil humus offers equal opportunity for the persistence or growth of pathogenic organism capable of causing crop disease. It is this aspect of soil microbiology which is of such importance to the plant pathologist, who is mainly concerned with the soil as a reservoir and source of plant pathogens. The presence or absence of such pathogens frequently bears little or no relationship to the rock structure of the soil; it is the character of the decaying vegetation present and the system of cropping followed which mainly determine their presence or absence.

B. Autonomous Dispersal in Soil

1. *The Contamination of Soil.* The original contamination of the soil with pathogens may occur in a variety of ways. It may take place by the gradual spread of the pathogen from a contaminate area to a non-contaminated area. or it may occur by the accidental introduction of contaminated material, such as crop debris or soil itself, into a clean area from one which is contaminated. The pathogen may also be introduced by the use of contaminated or infected seeds, or planting stock.

2. *Spread and Build-up of the Pathogen.* When once the soil has become contaminated, the spread and build-up of the pathogen will be determined by a number of interacting factors. The most obvious is the frequency with which as susceptible crop is grown. The nature of modern farming and cropping has frequently been such as to encourage systems of intensive cropping which tend to the production of a soil rich in plant pathogens capable of causing disease in epidemic form. Commonplace examples of such practice are world-wide. Spread by the mechanical transfer of soil is also of importance, but the spread and build-up of the pathogen will also depend upon its adaptation to the soil, and for this purpose soil-borne pathogens may be conveniently grouped into the three categories of nonspecialized facultative parasites, specialized facultative parasites, and obligate parasites.

a. Nonspecialized Facultative Parasites. These pathogens are capable of a complete saprophytic existence, but they possess the faculty of attacking live tissues and behaving pathogenically. They are considered as nonspecialist in that they can spread and build up in soil interpreted in the widest sense. Typical examples are *Pythium ultimum*, a common cause of the damping-off of seedlings as well as of other crop diseases; *Rhizoctonia solani*, which causes stem canker and black scurf of the potato and diseases of a great variety of host plants, and *Pectobacterium carotovorum*, the cause of soft rot of a variety of root vegetables. Both *Pythium ultimum* and *R. Solani* will grow readily in a typical cultivated soil, and when once the soil has become contaminated, the pattern of mycelial spread resembles that of an ever-widening circle such as is made by throwing a stone into a pond or by a fairy ring in a lawn. The growth in soil of *P. ultimum* has been studied by Smith (1954) and that of *R solani* by Blair (1943). The rate of spread and build-up of the organism may depend to some extent upon the original food base supplied, but it also depends upon the general nutritive value of the soil. Blair found that in the case of pure sand the growth of *R. solani* is greatly diminished after the removal of the food base, although in the case of the three samples of soil which he used, the diminution of growth was negligible in two of them but noticeable in the third after the food base had been removed. Factors such as soil temperature, aeration, and microbial antagonism also influence the rate of build-up and spread. Not much information is available regarding the spread and build-up of *Pectobacterium carotovorum* in soil although, according to Dowson (1957), the pathogen appears to live indefinitely in patches of inadequately drained soil. It was demonstrated by Kerr (1953) to be present in twenty separate Scottish soils.

b. Specialized Facultative Parasites. Specialized facultative parasites are those which, although capable of a complete saprophytic existence, depend in large measure upon the crop detritus of the host upon which they are pathogenic. They will not grow and build up in soil itself irrespective of its humus content. Some typical examples are *Armillaria mellea*, the cause of tree root diseases; *Ophiobolus graminis*, the cause of the take-all disease of wheat; and *Phymatotrichum omnivorum*, the cause of roct rot of cotton. These pathogens will not grow and ramify through the soil as do the nonspecialized types but require a food base such as the roots or crop debris of the host plants which they attack. Their spread and build-up in the soil is, therefore, closely linked with the frequency of growing susceptible crops. Taubenhaus and Ezekiel (1930) found that no spread of the mycelium of *Phymatotrichum omnivorum* occurred away from the roots of cotton, and they could produce no evidence for the independent growth of the fungus in the soil. At the same time it has been found that the fungus exists on some of the native plants of the uncultivated desert and spreads by the movement of infected roots from higher to lower lands by soil erosion and drainage. It has also been shown that *Ophiobolus graminis* only grows through the soil in association with the roots of its cereal and grass hosts. In the case of *Armillaria meliea* the growth of the fungus in space is extended through its ability to produce the typical rhizomorphs which grow out from the food base in infected wood. As examples of this, rhizomorphs were found to extend for 22 yd. from an infected pit prop in a mine-working by Ellis (1929), and Findlay (1951) found two rhizomorphs systems of this fungus extending for 10 ft. and 30 ft., respectively, in a water tunnel leading out of a reservoir 200 ft. below ground level; the food base for these was believed to be the timber used in the construction of the tunnel. Apart from adaptations such as this, however, the range of the specialized facultative parasite is strictly limited to the range of the host plant, and the pattern of soil contamination will largely conform with that for host cropping. The power of infectibility may not remain unimpaired in cases of rhizomorph production, such as that of *Armillaria mellea*, for Garrett (1956) states that "the radius of spread of rhizomorph systems from a food base is often much greater than the effective radius for successful infection." This is supported by Wallace (1935), who observed that infected roots of coffee and tea in Tanganyika are found only in close association with dead bushes. In considering automonous dispersal, however, the question of inoculum potential cannot be given full weight since the transfer of a pathogen in a viable state to a new site could bring about the raising of the potential necessary to procure successful infection. Conditions of soil and climate will also affect autonomous dispersal, and in this connection the work of Ezekiel (1940) with *Phymatotrichum omnivorum* is important. Ezekiel defines the cotton belt affected with root rot in the United States to be "from a point in southern Utah southwest into Mexico (where the southern limit is yet to be established) and from the eastern margin of Texas westward to southern California." Taking the isotherm delimiting the area where the lowest air temperature observed between 1899 and 1938 was –23° C., he points out that this follows rather accurately the northern limit of the occurrence of root rot from Arkansas west to Nevada. He further suggests that "root rot has persisted where the temperature has not fallen below –23° C., where the annual mean temperature has been 15.6° C. or higher, and where the frost- free period has averaged at least 200 days per year." Little alarm is expressed for the significant advance of the pathogen either by growth through the soil, which he estimates as 1 mile in 100 years, or by the more rapid introduction of the roots of diseased plants into the soil. McNamara (1926), working at Greenville, Texas, recorded an annual spread of about 10 ft. during the years 1921–1924. To the extreme west and east of the line denoted above, temperature does not appear to be the limiting factor. Taubenhaus and Ezekiel (1936) observe that the pathogen is favored by dark alkaline soils.

c. Obligate Parasites. The nature of the obligate fungal parasite is such that its existence is confined to the host plant which it attacks, and when the host plant dies, it can only survive in the soil in a resting state until reactivated by the renewed growth of the host. An obligate parasite, therefore, cannot grow and spread in the soil, its range and pattern will ideally be confined to the range and pattern of its host, discounting the mechanical transfer of contaminated soil. The rate and extent of the build-up of the pathogen will depend very much upon the frequency with which a susceptible crop is grown. Typical examples of soil-borne obligate parasites are *Plasmodiophora brassicae*, the cause of clubroot disease of Brassicae; *Synchytrium endobioticum*, the cause of potato wart disease; and *Spongospora subterranea*, the cause of powdery scab of potato. The only period during which obligate parasitic fungi could affect the spread of the pathogen in soil is when a sporangium or a resting sporangium germinates to liberate free swimming zoospores. The range of such zoospores, however, and the length of their active life are regarded as too limited to bring about any significant spread of the pathogen in the soil. Furthermore, when a zoospore is unable to contact and infect a susceptible host, it dies and the problem of soil contamination disappears. With regard to *Heterodera rostochiensis*, the nematode responsible for potato sickness, the larvae hatching from cysts will, according to Fuchs (1911), travel a distance of from 3 to 4 meters, while Baunacke (1922) records a distance of just over 9 meters. Here the spread of the pathogen in any year will also depend upon the number of generations passed and the soil type. The extent to which the pathogen is suitable for adaptation to new soil and climatic conditions to which it may be introduced are also factors of importance in determining the measure of its build-up and spread. Evidence is available indicating that when potato tubers severely affected with *Spongospora subterranea* were introduced for 3 successive years from a country where powdery scab is common and grown in another where the climate is warm and dry and yet where powdery scab is a prohibited disease, the occurrence of powdery scab on the crops grown from diseased tubers was never recorded in the country of introduction for the whole 3-year period.

The number of plant-pathogenic viruses reported as soil-borne is, so far, comparatively small. Some typical well-known examples are those causing tobacco mosaic, tobacco necrosis, wheat mosaic, and big vein of lettuce. There is no doubt, however, that soil transmission of plant viruses is more common than has hitherto been supposed and the number of known examples will be greatly increased as the study of this comparatively new branch of plant virology receives the attention it merits. Recently Cadman (1956) and Harrison (1956, 1958) have reported on soil-borne viruses responsible for disease of the raspberry in Scotland, while Walkinshaw and Larson (1958) have found a soil-borne virus associated with corky ring spot disease of potato in the United States. In so far as the active virus is associated with the cells of the host in which it occurs, the pattern of soil contamination will follow that of the distribution of host roots or detritus in the soil. Growing evidence supports the view that weeds play an important part in determining the incidence, spread, and survival of plant viruses so that the interpretation of "host" must be made in the wide sense to include such weeds as well as the crop affected.

3. *Persistence of the Pathogen.* The persistence of the pathogen in soil may vary from a few hours to many years; its effect on autonomous dispersal will closely depend upon the time elapsing between soil contamination and soil dispersal. Potato tubers may become infected with blight, caused by *Phytophthora infestans*, through the contamination of the soil with sporangia falling from the diseased haulms (stems). But the sporangia are short-lived and are mainly operative only in the top layer of the soil. Murphy and McKay (1925) have shown that from 2 to 3 weeks is the longest period that the fungus may be expected to survive and cause tuber infection and that such infection mainly occurs in the top 5 cm. of the soil. *Phytophthora infestans* will not survive for long periods in normal unsterilized soil away from its host plant.

Given continuously ideal growing conditions, there is no reason to suppose that nonspecialized facultative parasites such as *Phytophthora ultimum* and *Rhizoctonia solani* should not contaminate the soil indefinitely without the stimulus of a susceptible crop. Doubtless, they do persist in the soil for long periods, and this persistence is assisted by the production of oospores, chlamydospores, and resting mycelia, all of which help to tide them over periods unfavourable for growth.

More detailed information is available for some of the better-known specialized facultative parasites, and Bliss (1951) has suggested that *Armillaria mellea* will persist for 6 or more years in infected citrus roots of more than 3 cm. in diameter in some California soils. There can be little doubt that the rhizomorphs of this fungus are persistent although very definite evidence as to the extent of their longevity is lacking. For *Ophiobolus graminis*, Garrett (1944) and Fellows (1941) have shown that conditions restricting microbiological activity in soil, such as low temperature, dryness, and poor aeration are more likely to support longevity of the fungus in host tissues than those such as high temperature, adequate moisture, and good aeration, which encourage the activity of the fungus as a pathogen. An ample supply of nitrogen also helps to prolong the survival of the fungus by enabling continued mycelial development to occur in infected tissue. The degree of persistence of *O. graminis* in soil will, therefore, depend upon the interaction of all these factors.

In the case of *Phymatotrichum omnivorum* the formation of sclerotia is an important; if not limiting, factor in determining the longevity of the fungus in soil. Taubenhaus and Ezekiel (1930) and Neal and Maclean (1931) found that the vegetative strands of *Phymatotrichum omnivorum* survive their host roots by only a few weeks, whereas Ratcliffe (1934), Rogers (1937), and Rea (1939) concluded that a 4-years' crop rotation is necessary to obtain satisfactory control of cotton root rot. Neal (1929) first reported the occurrence of sclerotia of the fungus in the field, and it is by their production that the pathogen is enabled to survive. Rogers (1942) showed that sclerotial formation occurs after the conclusion of the parasitic phase of the pathogen. He found the greatest number of viable sclerotia after the first year's growth of a nonsusceptible crop following cotton. Taubenhaus and Ezekiel (1936) found the viability of sclerotia to range between 10 and 2% after 5 years in soil of medium moisture content. Ezekiel (1940) further recorded 10% germination after 6 years, 8% after 7 years, but found no viable sclerotia after 9 years.

Since the obligate parasite is wholly dependent for existence upon the living host, its persistence in soil away from its host must depend upon the production of resting spores or some such like-resistant structure. *Plasmodiophora brassicae* survives as resting spores, and 7 years seems to be a commonly accepted time for the survival of such spores. Gibbs (1939) has recorded 5 years for New Zealand, and Fedorintchik (1935) obtained infection in cabbage seedlings in land not sown to a crucifer crop for 7 years in Russia. Germination of a high proportion of the spores is likely to occur without much delay in fallow soil without the stimulus of the root of a susceptible host; this is known as "spontaneous germination". Spontaneous germination is more marked in acid soils of high moisture content than in dry alkaline soils in which the spores appear to survive longest. At the same time the marked reduction in the viable spore load in necessary to soils is offset by the fact that a much lower spore load in wet acid produce clubroot in epidemic form under such conditions than when the soil is dry and alkaline.

The persistence period for *Synchytrium endobioticum* appears to be even longer. In Germany, Schaffnit (1922) suggests at least 10 years in land fallowed and kept free from weeds. In Northern Ireland a susceptible potato variety contracted wart disease in 1956 when planted in land where the disease had first been recorded in 1926; very few, if any, crops of immune potato varieties only had been grown in this land during the interim period. It would thus seem that the resting sporangia of *Synchytrium endobioticum* might survive for periods of more than 10 years and even for as long as 30 years. Little direct evidence is available for the persistence in the soil of spore balls of *Spongospora subterranea* although they are frequently recorded as retaining their viability for several years. Some 6 to 8 years is the period usually accepted for the cysts of the nematode *Heterodera rostochiensis*. Thorne (1923) claims that after 6 years most of the cysts are empty or contain but very few eggs which, for the most part, are dead.

Of the soil-borne plant pathogenic viruses, tobacco mosaic virus, which is relatively stable, will survive in crop detritus and provide a source of infection for both tobacco and tomato crops over a period of at least 2 years. Wheat mosaic virus has been recorded as persisting in the soil in an active condition, with no wheat crop intervening, for from 6 to 9 years, and it has been suggested that it does so in some other soil-inhabiting organism which acts as a vector. Since the disease may be controlled by a insecticide, F. Johnson (1945) concludes that this acts on the vector rather than on the virus and suggested that the vector may be a nematode. The virus responsible for big vein of lettuce will survive in soil for more than a year, and the incidence of crop disease it will cause has been shown to be only slightly reduced when contaminated soil is diluted 800 times with sterilized soil.

C. Autonomous Dispersal by Soil

1. *By Soil Alone*. During the mechanical operations of cropping and farming, the movement of soil by plowing, harrowing, farm wagons and trucks, etc., will achieve the dispersal of a pathogen over the farm land. It will also be spread by soil carried on the boots of farm workers and by any soil erosion which may occur. The pattern of such spread will tend to be erratic and will fit in approximately with the pattern of farm activities. Through inter-farm communication such as the loaning of farm equipment by one farmer to another, and by land erosion on a larger scale, the pathogen may be spread from farm to farm and the pattern of such spread will again be erratic. In this way the spread may gradually increase until a whole village, a township, or province may be involved. Dispersal will also be achieved by the movement of soil in bulk, but as such a transfer ends to be uncommon in farm practice, this mode of dispersal will be more likely to occur in operations involving building or amateur gardening. The pattern of dispersal here will be even more erratic and much greater distances may be involved.

2. *By Soil Adhering to Plant Parts.* The transfer of contaminated soil from one area to another by its adherence to roots, tubers, and other plant parts is a most effective method of dispersal of a pathogen. It is in this way that pathogens may be spread not only from field to field and from farm to farm but from country to country and continent to continent.

D. The Prevention of Autonomous Dispersal by Soil

1. *Testing for Soil Contamination.* If soil is to be dispersed, whether avoidably or unavoidably, the knowledge of its freedom from or contamination with pathogens cannot be other than worthwhile. This emphasizes the need for techniques of soil examination for determining the presence of plant pathogens. The devising of suitable techniques which will give speedy, reliable, and accurate results is, however, not a simple matter, especially in the case of microorganisms and viruses whose minuteness makes difficult their direct visual detection in soil. Some success has attended the devising of such methods in the case of parasitic plant nematodes, and the best example is provided by *Heterodera rostochiensis*, the potato root eelworm, which is carried in the soil in cyst form. In many potato-growing countries soil surveys are in progress and are being made with the objects of defining contaminated and noncontaminated areas. Methods are also available for determining soil contamination with *Plasmodiophora brassicae*. These are the seedling host tests devised by Samuel and Garrett (1945) and Macfarlane (1952), and the pot method devised by Colhoun (1957) whereby, as the result of epidemiological studies, susceptible plants can be grown under conditions which will produce the disease in epidemic form. These techniques provide facilities for differentiating between clean and contaminated soil and allow for advice to be given, not only concerning the dispersal of soil, but also in regard to growing susceptible crops. The ultimate value of soil health surveys will be determined when it is known how control measures, adopted as the result of the knowledge gained, have been successful in preventing the spread of the pathogen and the preservation of noncontaminated areas clean and free from infection. For soil- borne plant pathogenic viruses, soil tests are at present largely limited to the growing of susceptible indicator plants.

2. *The Clearing of Contaminated Soil.* Another approach to the problem of preventing dispersal is by ridding the soil of the pathogen and thereby clearing contaminated areas. The outstanding example of this

method is the partial sterilization of soil by heat, steam, or chemicals whereby all pathogenic organisms are killed.

Soil treatment can be practicably employed only for relatively small land areas such as those in use for glasshouse crops or nurseries where the value of the crop is such as to warrant the expense of the operation. It has, however, also been used in cases of new outbreaks of infection with potato wart disease in the United States where the focus of contamination is small and where the stamping out of the pathogen can be justified by the prevention of its spread to a wider area. For this purpose Hartman recommends the use of finely pulverized copper sulphate at 2,500 lb. per acre, and claims that by the end of 1950 the disease had been eradicated from 440 infected gardens. Partial sterilization of the soil, using steam or formaldehyde, has become commonplace in the growing of nursery and glasshouse crops in areas where it is difficult or impossible to bring in fresh supplies of noncontaminated soil, and where the build-up of soil pathogens through intensive cropping is so marked as to render impossible the growing of healthy crops without soil treatment.

An alternative method of clearing contaminated soil is by following the land and controlling susceptible weeds. Where immune varieties are available, the growing of these allows for the starving out of the pathogen without having to forego the cultivation of the crop. This method has proved invaluable in the case of potato wart disease where the use of so-called immune varieties had solved a difficult problem in countries where the potato is a major crop. In these countries the tendency for their increasing use grows steadily, and in the United Kingdom no new variety of potato can be registered unless it is immune from attack. In Northern Ireland the potato acreage planted with immune varieties has increased from 69% in 1923 to 92% in 1957. The occurrence of new physiologic races of *Synchytrium endobioticum*, capable of attacking potato varieties hitherto believed to be immune, is an added complication and a development over which careful watch will need to be kept.

The use of decoy crops which stimulate the germination of resting spores but do not contract disease with the resultant production of resting spores, such as occurs in the case of *Matthiola incana* and *Lolium perenne* when grown in soil contaminated with *Plasmodiophora brassicae* is an interesting new line of investigation.

Many of the major problems of soil contamination have arisen because of the conscious or unconscious neglect of the principles of good crop husbandry, and much of the harm which has been done could undoubtedly be rectified by more cautious cropping and by reverting to a sounder form of the husbandry. This has already been realized in take-all disease of what, clubroot of Brassicae, and root rot of cotton, where epidemiological studies have brought valuable knowledge to bear upon the incidence of these diseases and how they may be best controlled.

3. *The Prevention of Soil Dispersal.* Where planting stocks and plant products are transported and where there is a risk of pathogens being dispersed with the soil adhering to them, it would not seem unreasonable to require, if at all practicable, that they should be freed from soil before dispatch. In some cases this principle has been adopted, an example being the transport of plants to the United States, where it is required that the roots shall be washed free from soil. The case of modern transport by rail, sea, and air with the increasing facilities for handling large bulks at greater speed, coupled with the remarkable increase in the volume of travel, has not eased the problem, and the danger of the autonomous dispersal of soil-borne pathogens in this way to new and uncontaminated areas is ever present. Perhaps no better example can be cited than that of the international trade in seed potatoes. In order to keep the potato crop free from virus infection a very considerable international trade in seed potatoes has developed whereby stocks of virus-free seed are imported regularly by countries with drier and warmer climates (unsuitable for virus-free seed production because of the prevalence of insect vectors) from countries where the climate is wetter and cooler and where stocks of such seed can be comparatively easily grown. The countries of northwestern Europe are developing an increasing trade with those of the Mediterranean basin and the continent of Africa; this trade is likely to continue until the importing countries have found it possible to grow their own healthy stocks. Although such seed may be relatively free from virus contamination, there is always, however, the risk of important such pathogens as *Synchytrium endobioticum, Streptomyces scabies, Spongospora subterranea, Phytophthora erythroseptica, Pectobacterium carotovorum, Heterodera rostochiensis*, and others through the medium of soil adhering to the tubers. Because of the lack of easily applied techniques for determining accurately the presence or absence of many of these pathogens in soil and the lack of knowledge of the epidemiology of the diseases they cause, phytosanitary regulation devised to prevent their introduction through the medium of soil have often had to be framed from the point of view of fear of the unknown rather than in the light of clear scientific knowledge. A more direct and

satisfactory solution of the problem of preventing dispersal through soil would be achieved if the principle of washing plants and plant products free from soil before dispatch could be generally applied. Difficulties inherent in putting this into practice become apparent when the need for washing, in an effective manner, large bulks of such produce as potato tubers or bulbs is considered. It is for this reason that a large scale mechanical plant for washing and disinfecting seed potatoes in Scotland presents an interesting new development. By means of this plant, the tubers are subjected to jet-pressure washing within 24 to 48 hours after lifting, and then, when completely freed from soil, to a process of disinfection before being quickly dried and stored. This development has been made possible by recent cumulative advances in mechanical engineering, and it shows how, with the facilities provided by an advance in one field of science, problems in another field, incapable of solution at the time, can be relatively easily solved later. One of the advantages claimed for such washed tubers is that they are free from contamination with the cysts of *H. rostochiensis* and if this is substantiated, a distinct advance will have been made which should appeal to all importing countries where the introduction of this pathogen is feared. Apart from this, the freeing of the tubers from soil also eliminates the risk of the introduction of soil-borne pathogens as a whole. The disinfection of the tubers after washing is not only an added precaution, but it also effects the control of tuber-borne pathogens, reference to which is made in the section dealing with autonomous dispersal by plants and plant.

II. SEEDS

A. General

Perhaps one of the major developments in plant pathology during the present century has been the realization of the extent to which the true seed may carry plant pathogens, and there is now no important agricultural or horticultural crop where the seed is not associated with one or more of the pathogen responsible for outbreaks of disease.

Seed-borne pathogens may be grouped into those which are carried with the seed and those which are carried on or in the seed. In the first group is included dodder (*Cuscuta trifolii*) occurring as seeds in a clover seed sample, flax rust, caused by *Melampsora lini*, affecting flax straw detritus in a sample of flax seed, ergot (*Claviceps purpurea*) occurring as sclerotia in samples of seeds of cereals and grasses, and the nematode (*Anguillulina tritici*) which occurs in ear cockles in samples of seed wheat. A typical example of a pathogen normally borne on the seed is bunt or stinking smut of wheat (*Tilletia caries*) which is carried as resting or brand spores adhering to the outside of the caryopsis. Fungal pathogens, which are carried on or in the seed are: *Helminthosporium* spp., the cause of seedling blight, root rot, leaf stripe, and leaf spot of cereals and grasses; *Fusarium* spp., the cause of seedling blight, foot rot, and ear blight of cereals and wilt of flax and other crops; *Phoma spp.*, the cause of black leg of beet and foot rot of flax; *Ascochyta pisi*, the cause of leaf, stem, and pod spot of pea, *Colletotrichum linicola* the cause of seedling blight of flax; *Polyspora lini*, the cause of stem break and browning of flax; and *Gloeotinia temulenta* the cause of blind seed disease of rye-grass.

Corynebacteium michiganese, the cause of bacterial tomato canker, is a typical example of a seed-borne pathogenic bacterium. The cause of loose smut of wheat, *Ustilago tritici*, exemplifies the type of seed-borne fungal pathogen where the contamination is carried as mycelium entirely within the tissues of the caryopsis. Few plant virus pathogens are known to be seed-borne, but those responsible for bean mosaic and lettuce mosaic are two typical examples.

B. Autonomous Dispersal by Seeds

The principles underlying the autonomous dispersal of plant pathogens by seeds have much in common with those enunciated for soil. But since the seed, in its dormant state, seldom provides suitable conditions of active growth such as may occur in soil, the pathogen is normally carried as a spore (resting or otherwise) or as mycelium (usually in a dormant state) within the tissues of the seed coat or in the tissues of the seed itself. It is clear that the dispersal of the pathogen will be closely correlated with the movement of seeds, and the pattern will tend to be more deliberate than for soil since there is not the same degree of accidental movement. For instance, soil erosion, the movement of farm machinery, and any communication between farms are all significant factors in bringing about the movement of contaminated soil whereas (although they cannot be completely ruled out) they are not so significant for seeds. There is a close relationship between the pattern of the seed trade and that of the autonomous dispersal of a seed-borne pathogen.

1. *The Contamination of Seeds*. Cases where the contamination of seeds is a simple mechanical mixture of the seeds of both host and pathogen are exemplified by dodder, where the seeds of *C. trifolii* occur

mixed with those of clover. In ergot of rye the open flowers of the host are infected with the ascospores of *C. purpurea*; the young ovary is attacked and invaded, and eventually becomes replaced by the typical horn-shaped sclerotium of the fungus. The sclerotia so formed become mechanically admixed with the unaffected seeds and are sown with them when putting in the crop. Ear cockles of wheat resulting from crop attack by the nematode *A. tritici* also become admixed with the grain and may be sown with the seed. The pathogen may be carried on the straw or crop detritus of a diseased crop, pieces of which may occur as an impurity in the seed sample. In flax rust, pieces of straw affected with the resting teleutospore stage of *M. lini* are commonly found mechanically admixed with the seed. The bunt balls of *T. caries* occur mixed with seed wheat, but they commonly become broken during harvesting and threshing, whereby the brand spores of the fungus are released to contaminate the grain; thus, this pathogen is normally carried as spores attached to the grain.

Seed contamination with pathogenic fungi and bacteria frequently takes the form of an invasion of the tissues of the seed coat or of the seed itself where it remains viable in a relatively quiescent state until the seed is sown. If the pathogen is capable of free sporulation, it may also be borne in the form of spores adhering to the seed coat. In these cases seed contamination may be translated to seed infection because, if contamination occurs in the early development stages of the seed and conditions are suitable, the pathogen attacks the seed, and the seed becomes diseased. In *P. lini*, for instance, which causes stem break and browning of flax, all stages may be found in a contaminated seed sample—from where the seed has been attacked and killed in an early stage of development, through seeds which are quite healthy but which carry the pathogen as mycelium in the outer mucilaginous layers of the seed coat, to those which bear only the conidia of the pathogen adhering to them. This condition applies equally well to many other seed borne pathogens included in this group, such as *Fusarium* spp., *G. temulenta, C. linicola*, and *A. pisi*. Contamination or infection of the seed in these cases is normally of external original, it being brought about by spores of the pathogen during the flowering, maturing, or harvesting of the crop. Pethybridge and Lafferty (1918) and Lafferty (1921) in the investigation of the flax pathogens *C. linicola* and *P. lini* traced the infection of the seed via to boll. The conidial of *Helmintho-sporium avanae*, the cause of seedling blight and leaf spot of oats, may be found attached to the pales, where they germinate to produce resting mycelia. Whereas the conidia of

Gibberella zeae attacking wheat may bring about flower infection or may attack through the glumes, the attack by other *Fusarium* spp. is normally confined to the glumes. In *A. pisi* the pea seeds contract contamination through the pod-spotting phase of the disease. *G. temulenta* infects ryegrass at the time of flowering through ascospores released from apothecia produced on infected seeds, while secondary infection is brought about by conidia produced in abundance by primary infection with ascospores. In this disease it is also easy to find every phase of contamination, ranging from seeds killed by the disease to those bearing conidia externally.

An interesting feature of a number of these diseases, such as those caused by *H. avenae, C. linicola*, and *p. lini*, is the healthy and vigorous growth of the crop, which usually occurs between the primary phase of the disease in the seedling stage and the secondary phase leading to the contamination or infection of the seed. Although it is generally assumed that the source of the inoculum for the secondary phase is supplied by spores produced by the primary and possibly intermediate phases, it is sometimes difficult to trace this in the field, and this fact suggests the need for further research. The diseases mentioned are largely specific, and the pathogens concerned are limited to one or a few kindred hosts. In the case of a nonspecific pathogen, however, such as *Botrytis cinerea* with a wide host range, seed contamination or infection may be contracted from outside sources irrespective of the state of contamination of the seed sample from which the crop is grown. Further contamination of the seed may spread rapidly, if weather conditions are suitable, in harvested and stooked crops of cereals, ryegrass, and flax in the case of such pathogens as some *Fusarium* spp., *G. temulenta* and *B. cinerea*.

Ustilago tritici, the cause of loose smut of wheat, presents a typical example of a systemic disease where the pathogen is borne is mycelial form entirely within the seed. Flower infection is affected by brand spores released from affected ears at the time of flowering of the crop, and the resultant grain produced may, according to the wheat variety, contain mycelia in the pericarp only or in all parts of the embryo and even in the root primordia.

The comparative rarity with which plant pathogenic viruses are transmitted by the true seed presents one of the most interesting problems of plant virology. Of the hundreds of plant virus diseases which have been described, only 45 have been reported as seed-transmitted. Reddick and Stewart (1919) first showed that bean mosaic virus is transmitted by the seeds of Phaseolus *vulgaris*. Later, Ainsworth and Ogilvie (1939) found that lettuce mosaic virus may be seed-transmitted and that some 5% of the seed used produced infected seedlings. The varying claims made for the seed transmission of tomato mosaic virus may be explained by the work of Ainsworth (1934), who showed that the virus occurs in the seed coats of seed grown from infected parents.

The ability of a virus to enter the seed depends upon the interaction of virus and host, for as Bawden (1950) points out, "In the same host, one virus may be seed-transmitted and another not, and the same, virus may be seed-transmitted in one host but not in others." Recent work by Crowley (1957), in which 5 mechanically transmitted viruses were used, indicates that host sterility will not account for the lack of seed transmission in any of the cases investigated. All 5 viruses infected the seeds of the hosts but only bean mosaic virus was found to infect the embryo. Crowley found no evidence for the production of virus in activators in seeds or developing embryos. He regards the rarity of seed transmission to be due to the inability of most viruses to infect megaspore or microspore mother cells of infected plants as well as to their inability to infect developing embryos because of the lack of plasmo-desmatal connection with the endosperm.

2. *Spread and Build-Up of the Pathogen*. The rate spread and build-up of the pathogen will vary according to the prevailing seasonal conditions. Given a series of successive seasons favorable for the growth and development of the pathogen in relation to the host, then a progressive build-up of the pathogen as a seed contaminant may be anticipated. The principle of seasonal conditions acting as a limiting factor controlling the rate and extent of seed contamination holds generally for seed-borne pathogens as a whole. If the host plant is regarded as a match box and the pathogen as a match, then seasonal conditions leading to seed contamination may be regarded as the act of striking the match. So long as the seasonal conditions are unsuitable, host and pathogen will tend to exist alongside each other; no outbreak of disease will occur, there will be no seed contamination, and the host plant will tend to flourish in spite of the pathogen. Whether there will be a gradual tendency towards the build-up of seed contamination over the seasons will depend, in the first place, upon the climatic conditions prevailing over the areas as a whole and, secondly, upon the specific climatic condition from season to season. This is illustrated by the seed health survey made for fiber flax from

1940 to 1955 in the United Kingdom, the results of which are discussed in the section dealing with the value of seed health surveys.

The pattern of the build-up of seed contamination in a crop may be even, but it may also be erratic and variable. Contamination may occur in pockets which have served as foci for the outbreak and spread of disease. This may be due to the uneven contamination of the seed sample sown or to the soil conditions favoring or disfavoring the outbreak and incidence of disease. That soil conditions of temperature and moisture can exert a very significant effect on the incidence of seedling blight of oats, caused by *H. avenae*, has been clearly demonstrated. When flax seed was sampled for the United Kingdom seed health survey, it was found that in bulks of seed which had not been thoroughly mixed, the incidence of contamination could vary greatly in different parts of the same bag of seed. An erratic pattern of seed contamination is also likely to occur when the pathogen is originally contracted from the soil, as in the case of *Phoma* sp., causing flax foot rot. Here the pathogen may be contracted in the first place from soil, and the attack becomes evident by the occurrence of odd plants here and there in the crop affected with the disease. When once the seed has become contaminated, the disease may break out in epidemic form in the crop if the seed is sown under suitable season conditions.

3. *Persistence of the Pathogen.* By persistence of the pathogen is meant the time during which the pathogen will remain viable with, on, or in the seed and assumedly capable of causing disease. It refers to the period during which the seed is stored rather than to the persistence of the pathogen in soil after the seed has been sown. Normally, the conditions required for the germination of the seed and the growth of the host plant will largely assimilate those for the pathogen, and it may be assumed that, after overwintering, the pathogen will become active soon after the seed is sown. There are instances, such as that of ergot, where the sclerotia may remain dormant if buried too deeply in the soil. In the northwest of the United States the brand spores of dwarf bunt of wheat, *T. caries*, may remain viable in the soil for up to 7 years. The time of persistence will clearly depend upon such factors as the nature of the pathogen itself, the condition in which it occurs as a contaminant, and the method of seed storage after harvesting the crop. In the case of a simple mechanically admixed contaminant, such as dodder, the problem is clearly one of comparative seed longevity, while for ergot it may depend upon the relative longevity of sclerotium and seed. In such cases

there is a dearth of very accurate knowledge of persistence time which, in any case, will vary with environmental conditions.

It is known that the nematode *A. tritici* in ear cockles of wheat kept dry will remain viable for up to 10 years. In herbarium material the brand spores of *T. caries* were found to be viable after 18 years and those of *Ustilago avenae* after 13 years. Although the brand spores of *Ustilago nuda* are relatively short-lived, barley seed infected with loose smut produced some smutted heads after storage for 11 years. *H. avenae*, which is normally carried as mycelia in the seed coats of oats, will retain its viability for a period of up to 10 years in Canada and for at least 7 in Northern Ireland. *Alternaria tenuis*, a mold commonly contaminating seed oats, tends to die out in the early years of the storage period, but with its death *H. avenae* is more readily isolated from an increasing percentage of seeds. It has been suggested that molds such as *A. tenuis* may behave antagonistically to *H. avenae*, thus interfering with its ready detection by growth until they have died out. The persistence of *Fusarium* spp. in seed oats ones not appear to be so marked because it has been found in Northern Ireland that both mycelia and spores tend to have lost their viability after seed storage for some 3 years. Seed-borne pathogens of flax may persist (from 1 to 6 years, as illustrated by the following examples; *C. linicola* (2 to 6 years), *P. lini* (1 to 5 years), *B. cinerea* (1 to 3 years) and *Phoma* sp. (2 to 3½ years). In blind seed disease of ryegrass, caused by *G. temulenta*, the surface-borne conidia lose their viability after 6 months, while the mycelia carried in the seed tissues will persist for up to 20 months. Generally speaking, surface-borne spores of pathogens, as might be expected, tend to lose their viability more rapidly than does mycelium, which is seated within the seed coat and seed tissues.

It may be concluded that the persistence of a seed-borne pathogen under normal conditions of seed storage will vary from a few months to as long as 10 years or more, according to its being surface-borne as thinwalled conidia, as chlamydospores, as sclerotia, as mycelium within the tissues of the seed coat or outer caryopsis, or internally as mycelium within the more deep-seated tissues of the seed. Little evidence is available in regard to the time of persistence of the virus in the case of the relatively fee viruses which are seed-transmitted although Middleton (1944) obtained as many virus infected squash plants from seed which was 3 years old as from freshly harvested seed. Working with a seed transmitted virus of muskmelon, Rader et. al. (1947) obtained 93 and 28% infected seedlings from the experimental lots of freshly harvested seed and only 3 and 6% after the seed had been stored for 3 years.

C. The Prevention of Autonomous Dispersal Seeds

1. *Testing for seed Contamination.* If autonomous dispersal of a pathogen by seed is to be prevented, then the possession of accurate knowledge as to whether and to what extent a seed sample is contaminated will be both desirable and helpful. Some indication as to the likelihood or not of the seed's being contaminated can often the determined by a system of crop inspection. If, for instance, loose smut is seen to occur in crops of wheat and barley, the assumption is that the seed saved from these will be contaminated with the loose smut pathogen. It can then be determined whether or not to use to seed for sowing or to ensure its treatment before sowing in order to control the disease. But crop inspection, although it is the only practicable method available in some cases, cannot be expected to give such a definite or accurate estimate as the direct examination of seed for pathogenic contamination. Methods for the testing of seeds for purity and germination have been evolved and practiced internationally for the last half century, but similar methods for the determination of seed health are of comparatively recent introduction, and those which are available are only just beginning to be used in general practice. This is due to the fact that until recently the need for determining seed health, which is some cases may be quite as important as purity and germination, was insufficiently realized and, at the same time, suitable techniques for carrying out the work were not available.

Notice is taken in routine seed testing of the presence of mechanically admixed pathogens such as dodder, ergot, bunt balls, ear cockles in wheat, etc., but this is more or less included in the purity analysis of the seed sample and cannot be considered as part of a real health test. Suitable techniques for the health testing of seeds must be capable of dealing as inexpensively as possible with large numbers of seed samples and should provide accurate and speedy results in no longer time than is required for carrying out a test for germination. This will allow for the results of purity, germination, and health testing to be made available at one and the same time. The aim of health test must be to provide an accurate estimate of the percentage of seeds in any given sample which is contaminated with the pathogen in a viable condition. In devising such tests attempts have frequently been made to use the methods employed in modern seed testing, with or without some modification, to cover the need for testing for health. But whereas this line of approach may on occasion provide some approximate information as to whether a particular pathogen is or is not present, or it may indicate in a general way whether the seed sample is healthy or unhealthy, it is not scientific and does not fulfil the conditions to be aimed at in health testing as state above.

During the period of the second world war thousands of tons of fiber flax seed produced in the United Kingdom were of the highest quality in so far as the standards of purity and germination were concerned, but a large proportion of the seed was very heavily contaminated with *C. linicola*. This was not and could not be revealed by the normal methods of seed testing and was only determined by use of a special technique devised for the purpose of testing for seed health. An approach to accurate testing for seed health has been made by using sterilized damp filter paper or agar medium upon which the seeds are planted out in Petri dishes and examined for the growth and development of fungal pathogens. More recently the use of filter paper has given way to the use of the agar medium method. The Ulster method, using a 2% malt extract agar medium, and the modified Ulster method, when the seeds are surface sterilized with a dilute solution of sodium or calcium hypochlorite before testing, have been found to give most reliable results in the case of a large variety of crop seeds and have recently been adopted as the standard international methods for the health testing of seeds of flax and oats. For flax the Ulster method is used for determining seed contamination with *C. linicola, P. lini, B. cinerea, Phoma* sp., *fusarium lini*, and *Alternaria linicola* as well as with molds in general, while for oats the modified Ulster method is used for determining seed contamination with *H. avenae, Fusarium* spp., and other molds.

For the contamination of ryegrass seed with *G. temulenta*, the water droplet method can be used for surface-borne conidia, but a modification of the Ulster method is necessary if the amount of internal infection is to be assessed. Headway has been made in developing techniques for the direct examination of seeds of barley and wheat for the presence of internal mycelia of the loose smut fungi, and this is exemplified by the whole embryo method. In any consideration of the problem of seed examination for contamination or infection with pathogenic fungi, due recognition must be given to the pioneer work of Doyer (1938) in the Netherlands.

2. *The Value of Seed Health Surveys.* When suitable techniques are available for the rapid and accurate assessment of seed health, it is possible to employ them not only for routine seed testing but also for carrying out planned seed health surveys for any particular district or

region. The value of such surveys is wider than the mere checking-up of seed health, for there is always the possibility of discovering areas where seed free from contamination may be grown or where contamination is so severe as to warrant the discontinuance of seed production. Such a health survey was carried out for fiber flax seed produced in the United Kingdom from 1940 to 1955. The Ulster method was used to examine more than 25,000 samples of seed drawn from production centers situated in the various counties of the United Kingdom and shown on the outline map (Fig. 3.1). As a result of this survey, extending over 15 years, it was shown that fiber flax seed produced in eastern and southeastern England is relatively free from contamination with *C. lini, P. lini,* and *Phoma* sp. and that there was no build-up of contamination in these areas. In the wetter and cooler regions west and north of the Pennines, conditions are more favorable for these pathogens, and seed contamination tends to build up and increase. Table 3.1 shows the survey results obtained for seed contamination with *C. lini.* Seed contamination with *B. cinerea* is general throughout the United Kingdom, with a tendency to be higher in upland and coastal areas. The incidence of *Fusarium lini* and *Alternaria linicola* in seed produced in the United Kingdom is negligible. If ever the need arises in the future for a planne program of flaxseed production in the United Kingdom, the results a obtained from this health survey indicate quite clearly that seed production should be confined to the east and southeast of England. Another result arising from this investigation indicates the possibility of clearing stocks of contaminated seed by growing the seed in areas where it is known that contamination of the progeny is not likely to occur. There can be little doubt that the extension of seed health surveys over wider areas and to include a range of agricultural and horticultural crops would furnish results of equal value to those obtained in the United Kingdom for fiber flaxseed and add substantially to the greater understanding of problems relating to seed health.

3. *Seed Disinfection.* In preventing the autonomous dispersal of seed-borne fungal and bacterial plant pathogens, seed disinfection is a most valuable tool, and with the introduction of modern disinfectants with increased facilities for carrying out the work on a large scale, much more is being done to prevent such dispersal. The introduction of organo-mercury compounds for general use with the seeds of cereals marked a significant advance, and there is now hardly a country in the world where the use of these disinfectant is not known and applied on a large or small

Fig. 3.1. Map showing localities of flaxseed producing areas, expressed as countries, in the United Kingdom, 1940–1955.

scale. The use of thiram as a disinfectant for the seeds of flax, peas, and beans, etc. marked a further advance in so far as this product is relatively nonpoisonous. The problem of seed disinfection will not be finally solved until a nonpoisonous product suitable for universal use has been discovered.

The hot water treatment for the internally-borne smut pathogens of wheat and barley is still efficacious, but the need for great care in

TABLE 3.1

Percentage of Seed Contaminated with *Collectotrichum linicola*

	1940	*1941*	*1942*	*1943*	*1944*	*1945*	*1946*	*1947*	*1948*	*1949*	*1950*	*1951*	*1952*	*1953*	*1954*	*1955*
England																
Suffolk	—	—	—	0.01	0	0	0	0	0	0	0	0	0	0	—	—
Kent	—	—	—	0	0.01	0	0	0	0	0	0	0	0	0	—	—
Norfolk	—	—	—	0	0.01	0	0.015	0	0	0	0	0	0	0	0.04	—
Cambridge	—	—	—	0.01	0	0	0	—	—	—	—	—	—	—	—	—
Sussex	—	—	—	0.12	0.01	0	0	0	0	0	0	0	0	0.01	0.3	—
Lincoln	—	—	—	0.01	0	0	0	0	0	0	0	0	0	0	0.23	0
Northampton	—	—	—	0.01	0.01	0	0	0	0	0	—	0.04	0	0	0.02	—
Yorkshire E.R.	—	—	—	0.01	0.02	0	0	0.03	0	0	0	0	0	0	0	0.01
Yorkshire N.R.	—	—	—	0	0.02	—	—	—	—	—	—	—	—	—	—	
Derbyshire	—	—	—	0	0.07	0	0.025	0	0	0	0	0	0	0	0	—
Wiltshire	—	—	—	0.11	0.02	0.37	0.32	0.41	0.13	0.01	0	0	0	0.12	0.7	0
Somerset	—	—	—	1.8	0.05	1.67	0.26	1.05	0.29	0	0.17	0.82	0	0.44	—	—
Dorset	—	—	—	0.16	0.24	2.5	—	—	—	—	—	—	—	—	—	—
Wales																
Pembroke	—	—	—	7.6	0.35	3.2	0.05	0.14	—	—	—	—	—	—	—	—
Scotland																
Aberdeen	—	—	—	1.7	7.5	8.1	—	—	—	—	—	—	—	—	—	—
Fire	—	—	—	3.3	0.19	0.05	—	—	—	—	—	—	—	—	—	—
Perth	—	—	—	2.6	2.2	1.1	—	—	—	—	—	—	—	—	—	—
Northern Ireland																
Down	1.6	11.7	7.1	4.7	3.2	0	0.21	1.6	0.6	0.13	0.8	—	—	—	—	—
Antrim	5.5	17.5	14.3	6.1	1.7	0.37	0.1	0.33	1.5	0	—	—	—	—	—	—
Armagh	—	20.8	16.3	8.0	4.0	0.1	—	—	—	—	—	—	—	—	—	—
Londonderry	5.4	13.3	26.1	4.0	—	—	—	—	—	—	—	—	—	—	—	—
Tyrone	2.5	17.7	18.1	8.4	4.7	1.7	0.29	0.72	0.81	—	—	—	—	—	—	—
Fermanagh	—	30.3	15.2	20.9	4.1	—	—	—	—	—	—	—	—	—	—	—

carrying it out works against its general use. Trends in recent research indicate that more attention is being given to the possibility of disinfecting seeds by physical means. This is to be welcomed, since, although the efficacy of some of the very poisonous chemicals now being used for plant protection is not doubted, equal efficacy from avoiding the use of poisonous substances would deserve serious consideration. Further, seed disinfection must not be regarded as a substitute for the production of healthy noncontaminated seed; it must not be used as a tool to make bad seed good but rather to make good seed better.

4. *The Prevention of Seed Dispersal.* Control by the prevention of seed dispersal appears to be the obvious answer to the problem if such action were possible or even desirable. The farmer himself frequently adopts this practice, for, if he suspects the seed he intends using to be contaminated and likely to produce a diseased crop, and if there is no other remedy, he will either destroy it or use it for some purpose other than sowing. Success may also attend the prohibition of sowing contaminated seed with the provision of some allowance for tolerance. When fiber flaxseed was being produced in the United Kingdom during the second world war, an epidemic outbreak of foot rot caused by seed-borne *Phoma* occurred in 1943. Since the routine method of seed disinfection with thiram (introduced in 1940) was found to be only about 50% effective for this pathogen, it was decided to discard all seed for sowing which was contaminated to the extent of more than 5%. This action was successful, and although seasonal variation may have played a part, evidence gained in the following seasons showed clearly that the principle of allowing only slightly contaminated seed which had been disinfected to be sown prevented any further serious outbreaks of this disease.

The storage of seed for a period before dispersal in order to allow the pathogen to die out has also suggested and itself as a control measure. It is sometimes practiced for celery seed affected with *Septoria apii*, the cause of leaf spot, in the belief that the pathogen is no longer viable in overyeared seed. Apart from the difficulty of getting conidia taken from such seed to germinate, no further evidence seems to be available on this point. Work with flax shows that seed-borne pathogens will die out after normal storage of the seed for some years. Seed of this age, however, will have lost some of its germination capacity and much of its energy of germination; so tuat this method of control cannot be recommended. The same argument holds for seed oats, where it takes at least 3 years for *Fusarium* sp. to die out and up to 10 years for *H. avenae*. All trace of viability of *G. temulenta* in ryegrass seed disappears after 20 months' storage—a shorter period, and, therefore, more hopeful.

Much can be done by preventing the dispersal of badly cleaned seed. This is exemplified not only in the case of dodder, ergot, wheat ear cockles, and flax rust but also where badly contaminated or infected seed is lighter and smaller and, therefore, capable of being taken out of the sample by more rigorous cleaning. Such is the case with *C. linicola* and *P. lini* of flax as well as with some other fungal contaminants. Middleton (1944) found in the case of virus diseases of squash that light seed produced up to 37% of infection and heavy seed only 2%; he states that most infected seeds can be removed by careful winnowing. All attempts to prevent the dispersal of fungal and bacterial pathogens by preventing seed dispersal must be given the broadest consideration.

Epidemiological studies have shown that climatic conditions may be a limiting factor in determining the spread and build-up of seed-borne pathogens, and it should be remembered that it is possible to disperse contaminated seed to districts and areas where growth conditions are such that they will lead to the eradication of contamination and to the harvesting of clean, noncontaminated seed.

III PLANTS AND PLANT PARTS

A. General

Plants and plant parts which are dispersed as planting stock or farm and garden produce and plant debris which may accumulate as waste in the course of cropping, etc. are considered under this heading. Crop debris and decaying vegetable matter which remain in the soil and help to build up its humus content have been dealt with under the heading of soil. Traffic in plants and parts of plants grows apace, and the problem of combating the autonomous dispersal of plant pathogens through the medium of planting material and plant products assumes growing complexity. Whether the commodity be complete plants, tubers, bulbs, corms, runners, cuttings, or produce for consumption, the need is that it shall be healthy and free from viable pathogens and, in particular, from those which are likely to cause disease or diseases unknown or unrecorded in the place of its introduction. It is the real danger of the spread of disease in this way which has occupied the minds of legislators and administrators, as well as plant pathologists, ever since it has been understood how disease epidemics can arise through the traffic in plants.

The introduction of potato blight (*Phytophothora infestans*) into Europe, with its resultant disastrous effect, as illustrated by the famine in Ireland from 1845 to 1847, is but one example, and the appearance in Europe circa 1900, of American gooseberry mildew, caused by *Sphaerotheca mors-uvae*, with its temporary retarding effect upon the cultivation of gooseberries, is but another. Since the problem of adhering soil has already been considered, the pathogens discussed here are those with which plants and plant parts themselves are likely to be infected.

B. Autonomous Dispersal by Plants and Plant Parts

1. *The Contamination of Plant and Plant Parts.* The contamination of plants and plant parts means their infection with pathogenic organisms. The rate and pattern of contamination will follow largely that of the disease with which the crop may become infected. Such contamination will, in most cases, be contracted during the growing of the crop although it may occur during storage, particularly in the case of plant produce.

2. *Spread and Build-Up of the Pathogen. a. Planting Stocks and Plant Products.* The spread and build-up of the pathogen will take its pattern from the season and the conditions of growth and storage. While the degree of its spread and build-up is of great importance, the actual presence of the pathogen is equally important. The fact that it is present at all in a viable condition will allow for its establishment and spread in a new area if the environment is congenial. If potatoes or gooseberries are to be moved for planting to a new area, where it is known that the diseases with which they are likely to be affected are likely to cause harm, then it is important that they should be free not only from mass infection but from all infection.

Accumulated knowledge confirming the existence and importance of physiologic races of pathogens such as *S. endobioticum*, *P. Brassicae*, and *Venturia inaequalis*, the cause of apple scab, indicate the danger of the introduction not only of the pathogen, but of a new physiologic race capable of causing severe crop disease. Although the transmission of pathogenic viruses has been shown to be relatively uncommon in the case of true seeds, the reverse holds for crops which are vegetatively propagated. The probability of complete systemic infection with the virus which results in the permeation of all the plant tissues means that infection will remain in that plant or clone as long as it is cultivated. The ready spread and build-up of pathogenic viruses in vegetatively propagated plant material is, therefore, of special significance and has led to the growing demand for planting stocks which are free from virus contamination. The spread and build-up of contamination in planting stocks and plant products ceases to be of interest when it has reached the stage where the host has been killed or rendered of such poor quality as to be worthless; it then becomes plant debris.

b. Plant Debris. Although plant debris may not be generally subject to the same extensive movement as planting stocks and plant products, its accumulation may be a significant factor in securing autonomous dispersal. This is of special local importance in spreading disease from field to field and from farm to farm. Heaps of discarded potato tubers affected with blight caused by *P. infestans* will, if the winter is mild, carry the pathogen through to the next season, when it will sporulate and serve as a source of inoculum for neighboring crops. Crop debris from Brassicae affected with clubroot (*P. brassicae*) may be distributed with manure, while *B. cinerea*, the fungus responsible for gray mold, becomes readily established in decaying plant parts. Prunings from perennial crops also constitute a serious danger and even fallen rose leaves affected with black spot, *Diplocarpon rosae*, ensure the dispersal and survival of the pathogen. As useful as the compost heap in a garden may be for providing a supply of organic manure, it nevertheless consists of a mass of decaying plant debris, and the danger of the dispersal of plant pathogens, in compost may be very real. The ready growth of nonspecialized facultative parasites, such as *P. ultimum* and *R. solani* in debris rich in leaf mold, is also likely to occur and must be reckoned with when such material is used for plant propagation.

3. *Persistence of the Pathogen. a. Planting Stock and Plant Products.* Since planting stocks are normally not stored for long periods and are meant to be used with the least possible delay, the ability of the pathogen to survive over the winter or a dormant period will determine its reaching the new planting site in a viable condition. In the case of transplants such as tomatoes, chrysanthemums, and Brassicae, the adaptability of the pathogen to persist is not of great importance because only a few hours or days at the most are likely to intervene before the stock is replanted in its new site. This means that, except in occasional cases, when the pathogen may be killed during the period of transit or storage, the risk of its dispersal is always present. Even in cases where a fungus or bacterium is carried as delicate mycelium or in a nonresting state within the tissues of the host, such as *Peronospora destructor*, the cause of

downy mildew of onion, or *Corynebacterium sepedonicum*, the cause of potato bacterial ring rot, the conditions of storage will normally allow for the survival of the pathogen. In most cases the pathogen will be present as resting mycelia, resting spores, or sclerotia, and thus adapted to withstand adverse conditions. Examples of the persistence of the pathogen as resting spores are provided by *P. brassicae, S. subterranea* and *S. endobioticum* as well as pathogenic bacteria which form such spores. The cleistocarps of *S. mors-uvae* will survive for long periods until conditions are suitable for the release of ascospores. *B. cinerea* and *R. solani* normally form resistant sclerotia which will retain their viability for years. *Streptomyces scabies*, the cause of common scab of potato, will certainly persist in the tuber from one growing season to another. Viruses are persistent in planting stocks and are readily carried over during the normal period of storage.

Plant products, although used for consumption or decoration, can also serve as agents for autonomous dispersal. Apple scab (*V. inaequalis*), potato bacterial ring rot (*C. sepedonicum*), and virus diseases are cited as examples where the pathogen may survive and persist on or in such produce.

b. Plant Debris. The condition for survival and persistence of the pathogen in planting stocks and plant products also hold largely for plant debris, which may for many years provide a source of inoculum. Perennial crops are cited as an example where fallen leaves and prunings enable a pathogen to persist until the next growing season when they may be instrumental in assuring dispersal. In apple scab and black spot of roses, fallen leaves may provide a potent source of seasonal inoculum while in apple canker, caused by *Nectria galligena*, and American gooseberry mildew the prunings from trees and bushes may behave similarly.

C. The Prevention of Autonomous Dispersal

1. *Inspecting and Testing for Infection and Contamination*. As with soil and seeds, reliable methods for testing plants and plant products for contamination or infection with a pathogen provide a valuable tool in helping to prevent its dispersal. In the case of material affected with fungal or bacterial diseases the task may be relatively simple since the symptoms of disease will normally be observed, and symptomatology, aided by examination for the presence of the pathogen, will provide the information required. It is not a difficult task to determine whether *Brassica* seedlings are affected with *P. brassicae*, gooseberry transplants

with *S. mors-uvae*, or seed potato tubers with *S. scabies, S. subterranea, S. endobioticum, R. solani*, or *Pe. carotovorum*, and an examination of the stock, supplemented, if possible, by field inspection and some knowledge of the growing crop will provide an accurate assessment of plant health. These are the methods normally employed in the granting of health certificates for the export of planting stocks and plant products. At the receiving end it is not uncommon for some form of plant quarantine to be set up whereby incoming planting stock can be grown on in isolation and under constant observation for the purpose of health determination.

The contamination of planting stocks with viruses presents a different and more difficult problem because there is frequently no outward sign to indicate whether or not the stock is affected. Symptoms present in growing stocks, such as transplants, may allow of some assessment being made, but in dormant stocks such as potato tubers, bulbs, corms, and leafless trees, etc., there is usually nothing to indicate virus infection. It is for this reason that inspection of the growing crop from which the stock is derived is practiced as a normal procedure for gauging the likely state of health of the produce. Systems of crop inspection have now come to be practiced on a world-wide scale. No better example can be cited than that of the seed potato crop where so much has been done to prevent the dispersal of crop pathogens by this method. But even so, the problem of determining health by crop inspection alone is made difficult by the varying symptom pictures presented by different potato varieties affected with the same virus, by infection with a virus complex, and by the existence of symptomless carrier varieties of potato.

With increasing knowledge and development of new techniques, however, it has been possible to elaborate testing methods whereby a more certain and accurate approach to the problem can be made. The importance of this direct approach is being rapidly appreciated, and these new techniques are being used on an ever-increasing scale to supplement or replace visual inspection in the field. Briefly, testing methods for virus contamination fall into the three categories of grafting, the sap inoculation of indicator plants, and serology. Although not so widely used for the more easily recognizable infections such as leaf roll and severe mosaic, they are now in common use for milder infections not readily determined by eye inspection. Tuber core grafting can be used for potato leaf roll, while stem grafting is employed for the determination of contamination with viruses X, A, B, and C; a scion of the stock to be tested is grafted onto a variety such as Craig's Defiance, which is field-immune to all 4 viruses. Sap inoculation from potato stocks of determinate health is made onto such indicator plants as *Nicotiana tabacum* (White Burley), *Nicotiana glutinosa, Datura stramonium, Solanum demissum* (C.P.C. 2103.1 for virus Y and P. 1.175404 for virus A), and *Gomphrena globosa* whereby characteristic symptom pictures of such sap-transmissible virus infections as are caused by viruses X, Y, and A are obtained. The introduction of serological methods marks yet another advance, since by using either the precipitin or agglutination methods, a rapid technique is made available for the detection of viruses X and S. Grafting techniques have also been evolved for use with stocks of strawberries and raspberries. Healthy wild strawberry plants (*Fragaria vesca*, if inarch grafted with a strawberry variety of unknown health, will act as indicators and provide a symptom picture of the virus carried. Similarly, by a modification of Harris's method, adapted to bottle grafting, and by using certain raspberry varieties as indicator plants, the health of raspberry stocks may be determined. The extended use of the electron microscope may prove a further boon to testing for health if, by further simplification, rapid and easily applied techniques become available for the direct observation of the virus. In preventing the dispersal of pathogenic plant viruses, testing methods are not only useful for the examination of planting stock if and when required, but are proving invaluable as routine methods for the conduct of health surveys and for use in the building up of foundation stock of planting materials known to be free from virus contamination.

2. *Decontamination. a. Planting Stocks and Plant Products*. The ideal method for preventing the dispersal of plant pathogens by plant and plant products is the obvious one of growing healthy crops. This may be achieved by cultivation in areas where conditions are unfavorable for the incidence of disease or by introducing plant protective measures against the onset of disease. But in spite of all that can be done there will be instances and seasons when planting stocks and produce will carry some kind of infection which will call for the application of some process of practicable decontamination. The form of treatment most commonly applied is chemotherapeutical and is exemplified by the dipping of young apple trees into Bordeaux mixture before export or distribution for the control of *V. inaequalis*, the steeping of raspberry canes into a weak solution of an organo-mercurial to control *Erwinia tumefaciens*, the cause of crown gall, the immersion of vine cuttings into a 25 to 30% solution of ferrous sulfate to control *Elsinoe ampelina*, the cause of vine anthracnose, and the steeping of bulbs of tulips and narcissi into a

solution of formaldehyde or a mercury salt to prevent the transmission of bulb-borne diseases.

The evidence suggests that some such treatments have been adopted as emergency measures without the undertaking of adequate experimental work, but there can be no doubt as to the need for further research on this problem. The disinfection of seed potato tubers for the control of tuber-borne fungal pathogens has received considerable attention, and the use of solutions of hot or cold formaldehyde, acidulated mercuric chloride, and organo-mercurials have all been recommended. The dipping of tubers into a solution of an organo mercurial within 24 hours after digging has being promising results for the control of late blight caused by *P. infestans*, dry rot caused by *Fusarium coeruleum*, common scab caused by *S. Scabies*, some forms of black scurf caused by *R. solani* and skin spot caused by Oospora pustulans but not of powdery scab caused by *S. subterranea.*

The large scale plant for washing and disinfecting seed potatoes in Scotland, to which reference has already been made, operates on the principle outlined above and aims at the production of high grade seed free from soil and viable tuber-borne pathogens. Although fumigation is commonly practiced for the decontamination of stocks affected with insect pests, it has not found favor in use on other pathogens. Heat therapy, applied as a hot water treatment, will free narcissus bulbs from infection with the nematode *Anguillulina dipsaci*. The bulbs are immersed in water for 3 hours at 110° F. It has also been used to free-mint (*Mentha* sp.) runners from rust caused by *Puccinia menthae* by immersing them for 10 minutes in mid-February in water at 105 to 115° F.

Mention is made of the ingenious theory advanced by Jensen (1887) to account for the introduction of the living blight fungus (*P. infestans*) into North America and Europe. Having found that the mycelium in potato tubers is killed at a temperature of 25° C., he concluded that during the first three centuries of potato cultivation the tubers were disinfected by heat in old sailing vessels while passing through the tropics. With the introduction of the steamship (*circa* 1830 to 1840) the time taken to pass through the tropics was much reduced—with the result that the tubers were not decontaminated, and the pathogen survived the voyage in a viable condition. Heat therapy has also been applied for planting stocks contaminated with viruses. Kunkel (1935) reported the recovery of peach trees from yellow disease when kept for 14 days or more at 35° C., while Kassanis (1949) freed potato tubers from the leaf roll virus by keeping them at 37° C. in a moist atmosphere for a period of from 10 to 30 days. Bawden (1950) suggests that this method of treatment is not likely to be suitable for general application, but, at the same time, its possible value for freeing small quantities of valuable planting stock from virus contamination must be kept in mind.

The risk of dispersal through plant products also exists—especially for such produce as potato tubers intended for consumption but which could be used for seed. It is, therefore, most desirable to ensure that incoming produce is used for the purpose intended and not allowed to be distributed willy-nilly through areas where it could be a danger to the health of cultivated crops.

b. Plant Debris. The destruction of plant debris, either by burning or treatment with some means to render innocuous any pathogens it may carry, is the obvious way of combating the spread of disease through the medium of plant refuse, etc. Waste produce from stores and markets, prunings from fruit bushes and trees, discarded tubers, and roots thrown out in the course of sorting or grading the crop all come under this heading. Reference has already been made to the danger of the compost heap as a reservoir of viable plant pathogens. In the case of perennial crops of trees and bushes the fallen leaves may provide a source of inoculum each year for the next season, and where it has been found impracticable to collect and remove them, some such treatment as the spraying of the orchard floor with a suitable fungicide for the control of apple scab or they heavy mulching of the soil for the control of black spot of roses has been recommended.

3. *The Prevention of Plant and Plant-Part Dispersal.* Where doubt exists as to the possible contamination of plant parts or products and where decontamination is impossible or impracticable, their dispersal in any area where the contamination is likely to become established and cause serious outbreaks of disease should be prevented. It is due to the failure to recognize or to realize the danger of dispersing contaminated material, either through sheer ignorance or through the lack of epidemiological knowledge, that the need has arisen for so much plant-protective legislation and the introduction of that most expensive and ideally unnecessary luxury, the plant quarantine.

Much of the early quarantine legislation had to be framed in the light of very incomplete knowledge, and like poor King Canute in this abortive attempt to persuade the flowing tide to ebb, the earlier legislators got their feet very wet. The Destructive Insects and Pests Act for England

and Wales became law on Independence Day, 1907. Its first task was to stem the tide of entry of American gooseberry mildew into the country, since England was believed to be almost free from the disease. Inspection, made possible through the act, showed the disease to be already common in some of the country's southern counties. For some time it has been very doubtful whether legislation framed to prevent the passage of a pathogen from one country to another can be effective when there is only a land border between the two, and with the increase in case of speed and travel the value of an intervening ocean would now seem to be almost just as doubtful. The complicated boundaries which have had to be evolved of necessity to preserve the political integrity of the strains and races of mankind do not appear to serve the same or any other purpose in the case of microorganisms.

Whereas, in the past, it has seemed the bounden duty of an importing country to prevent the inlet of produce which might be suspect, with increasing knowledge the emphasis has shifted to the exporting country, which should prevent the export of such produce. In fact the time should come and may come (it has come in the case of seed potatoes) when exporting countries will vie with each other for the good health of their products. It is much easier and more sensible for a producing country to vouch for the health of its produce than to place the onus for determining health onto the expensive inspectorate of an importing country, which very often can only say that the incoming material is free from all disease and contamination to the best of its belief.

Nor must the introduction of the pathogen as such be regarded as the bogey incarnate unless there is a grave risk of its introduction's causing serious crop disease. Reference has already been made to the possible desirability of sending stocks of contaminated planting material to be grown in areas likely to clear their contamination. The inhabitants of Switzerland do not object to persons sick with tuberculosis entering their country to effect a cure; the climate of Switzerland encourages the recovery of the host and discourages the activity of the pathogen. The immediate need is for less legislation on a small national scale and for more research into epidemiological problems and disease control, with the object of putting into practice the knowledge so gained. The findings of science must be treated as a whole; they must not be taken piecemeal and prostituted for political, financial, or national ends. It is only in this way that one of the objects of the Second International Plant Protection Convention of 1951 can be attained. That object is that all plant-protective legislation should be framed with a view to encouraging international trade, and not to hindering such trade. When more research has been achieved and the accrued knowledge allows for a more truthful interpretation of the facts, it is essential that the findings be safeguarded and that their correct interpretation be preserved. If this cannot be done, the last state of things may be worse than the first.

4

DISPERSAL BY AIR AND WATER

PART A

The dispersal story of a fungus can usually be divided into three major episodes: first, liberation of spores or their actual escape from immediate contact with the parent tissue; secondly, their dispersal in a viable condition to a greater or lesser distance; and thirdly, the coming to rest of the spores on solid substrata,on some of which germination and successful establishment may occur. This chapter is concerned essentially with the first episode although the second cannot completely be ignored, because the efficiency of take-off can be judged only in relation to subsequent dispersal. In particular it is important to consider the turbulence or nonturbulence of the air into which spores are liberated. Before discussing this matter, however, the various types of spore liberation will be considered.

This account will be largely concerned with organisms of importance in plant pathology although reference will often be made to saprophytic types in which particular dispersal mechanisms have been more fully studied than in essentially similar pathogenic forms. However, certain examples, highly interesting in the general context of dispersal in fungi but of no phytopathological importance, will be ignored. Thus the beautiful discharge mechanisms of *Pilobolus, Ascobolus*, and *Coprinus* spp. will receive no mention, nor will the wide range of dispersal types displayed in Gasteromycetes be considered.

The conspicuous part of a fungus is essentially concerned with the production and liberation of spores, the feeding part being usually hidden away as a branched mycelium in the nutritive substratum. If we hope to understand a reproductive structure in a fungus, the question must be asked: "How are the spores set free?" It is extraordinary how often, even for the commonest fungi, no really satisfactory answer can be given to this question.

It is convenient to recognize two contrasting types of spore liberation. In the first, the spores are actively and violently discharged. In the second, liberation is passive in the sense that the energy concerned comes from outside the fungus, the dislodgment of the spores being due to the kinetic energy of wind or rain.

I. VIOLENT SPORE DISCHARGE

In connection with the violent discharge of projectiles the size of fungal spores (mostly less than 50μ in diameter) or small spore groups, certain basic mathematical expressions should be considered.

According to Stokes' Law, the rate of fall of a minute spherical particle in a fluid is given by

$$V = \frac{2\rho - \delta}{9\mu} ga^2$$

where V = the terminal velocity of fall in the medium

ρ = the density of the falling sphere

δ = the density of the fluid

g = the acceleration due to gravity (981 cm./sec.)

μ = the viscosity of the medium (1.8×10^{-4} in the case of air).

This can be simplified to

$$V = \frac{2ga^2}{9\mu} \tag{1}$$

since for fungal spores the density is approximately 1.0, and since the density of the air can be neglected. Turning to the question of the horizontal discharge of a spore we find, according to Barlow that

$$H = \frac{gD}{V} \tag{2}$$

where H is the initial horizontal velocity of discharge and D is the distance of horizontal throw. Combining (1) and (2), we get

$$D = H\left(\frac{2a^2}{9\mu}\right) \tag{3}$$

Thus for a given initial velocity of discharge, the distance of horizontal throw is proportional to the square of the radius of the spherical projectile.

The path of a spore projected horizontally is

$$y = \frac{V^2}{g}\left[-\log_e\left(1 - \frac{x}{D}\right) - \frac{x}{D}\right] \qquad (4)$$

y being the vertical distance of a point on the curve below the level of the point of departure, and x the horizontal distance from the vertical axis through this point. This curve (Fig. 4.1) has been referred to by Buller (1909) as a "sporabola." It is of interest to compare distances of

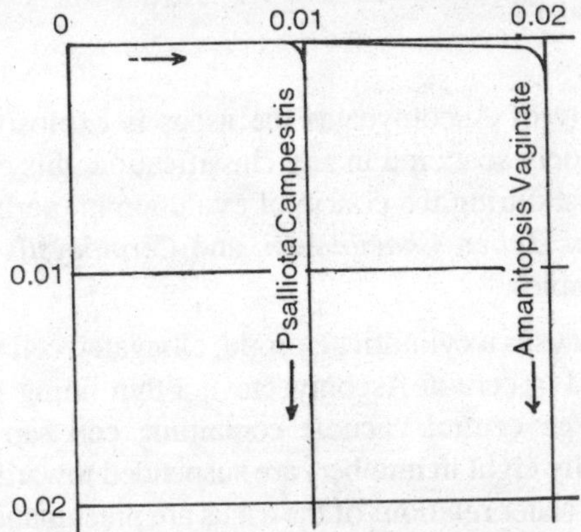

Fig. 4.1. The sporabolas of two spores shot horizontally from the hymenium. The spores, drawn to scale, are shown below. The scale is in centimeters.

horizontal discharge (D) and of vertical upward discharge (U):

$$U = D - \frac{g}{k^2}\log_e\left(1 + \frac{k^2 D}{g}\right] \qquad (5)$$

k being given by $9\mu/2a^2$

For particles the size of fungal spores U is almost as large as D. For example, in the common coprophilous pyrenomycete *Sordaria fimicola*, the spore projectiles are relatively large and an eight-spored mass may be shot horizontally to a distance of 10 cm. This projectile consists of eight spores, each $22 \times 13\mu$ and surrounded by a sheath of mucilage 3μ wide. The total mass has a radius of 21.6μ. Substituting in (5) it is found that $U = 9.82$ cm. In fact it has not been found possible experimentally to demonstrate a consistent difference between the distance of horizontal (10.0 cm.) and vertical discharge. The difference is still less where smaller projectiles, those normally encountered in plant pathology, are concerned.

Violent discharge of the characteristic or "perfect" spores is the rule in two great groups of fungi, Ascomycetes and Basidiomycetes, although there are many individual examples in both groups in which discharge of ascospores or basidiospores is no longer active. Outside these two groups there are few examples of violent discharge, for in the Phycomycetes generally and in nearly all conidial fungi, including the conidial stages of Ascomycetes and the vast hordes of Fungi Imperfecti, liberation of spores is essentially passive. In this chapter violent discharge will be considered first.

A. Ascomycetes

In the majority of Ascomycetes the ascus is explosive although in many genera, widely scattered in any classification, this explosive character has been lost during the course of evolution or, perhaps, has never existed. *Eurotium, Tuber, Chaetomium,* and *Ceratocystis (Ophiostoma)* are familiar examples.

The typical ascus is a cylindrical, turgid, elongated cell with a thin cell wall (two-layered in certain Ascomycetes), a thin lining layer of protoplasm, and a large central vacuole containing cell sap in which the ascospores (usually eight in number) are suspended toward the upper end of the ascus. The water relations of the ascus are presumably like those of most other living cells.

During the later stages of maturation the glycogen reserve (staining chestnut brown in iodine) disappears and is probably converted into sugar, which raises the osmotic pressure of the ascus sap. Unfortunately, careful plasmolytic determinations of the osmotic relations of maturing asci have not been made, nor is there any critical information about the osmotic pressure in ripe asci.

The ascus eventually bursts in a definite matter, most frequently either by the flinging back of a small apical lid (operculate Ascomycetes) or by changes at the apex, producing a minute pore (inoperculate Ascomycetes). Depending on the size and form of the apical opening in relation to the size and form of the spores, the latter are discharged simultaneously (or apparently so) or in obvious succession.

Compared with the basidium, which can rarely throw a spore more than 0.2 mm. and never more than 1 mm., the range of the ascus is great—being rarely less than 1 mm., usually of the order of 5 to 10 mm., and sometimes (e.g., in *Ascobolus immersus* and *Pleurage fimiseda*) as great as 500 mm. This very much affects Ascomycetes in relation to liberation of spores into turbulent air. Close to the ground or close to a host surface there is normally a layer of almost still, nonturbulent air commonly of the order of 1.0 mm. thick. Most Ascomycetes are capable of shooting their spores through this laminar layer into the turbulent air beyond.

The structure of the fruit body in Ascomycetes in relation to violent spore discharge varies. Three major types can be recognized: (1) the Discomycetes type, in which the discharge occurs from extensive exposed hymenia; the Pyrenomycetes type, in which the asci are enclosed within a flask-shaped, true perithecium, or in a biologically similar pseudothecium, where the asci must elongate singly up the neck canal to the ostiole before discharge; and the Erysiphales type, in which the asci are completely enclosed within a cleistothecium, the wall of which must first be ruptured before an ascus can emerge to scatter its spores. These types will now be considered separately.

1. *Discomycetes Types*. The organization of the apothecium in relation to spore discharge has been considered in some detail by Buller (1933). Although the examples he analysed are not directly the concern of the plant pathologist, the principles of organization are fully applicable to genera of phytopathological importance such as *Sclerotinia, Phialea*, and *Trichoscyphella*.

One example, *Aleuria vesiculosa* (Fig. 4.2), a fairly common species on dunged soil, will be considered. The apothecium is cup-shaped and usually several centimeters in diameter, with a palisade-like lining layer of asci at various stages of development intermixed with paraphyses. When an ascus bursts, an apical lid hinges back, the ascus wall contracts longitudinally and laterally, and the ascospores are shot to a distance of several centimeters. If contents of the asci were discharged at right angles to the hymenium, it is clear that many spores would simply be shot onto an opposite hymenial surface. This would be particularly true of those species with apothecia of a more vase-shaped form. However, this type of wastage does not occur, since the spores are shot freely into the air above the apothecium because of the phototropism of the individual ascus.

A special feature of most apothecia is the phenomenon of puffing. During a period of quiescence many asci ripen, but remain in a condition of unstable equilibrium. In this state a touch or perhaps a sudden change

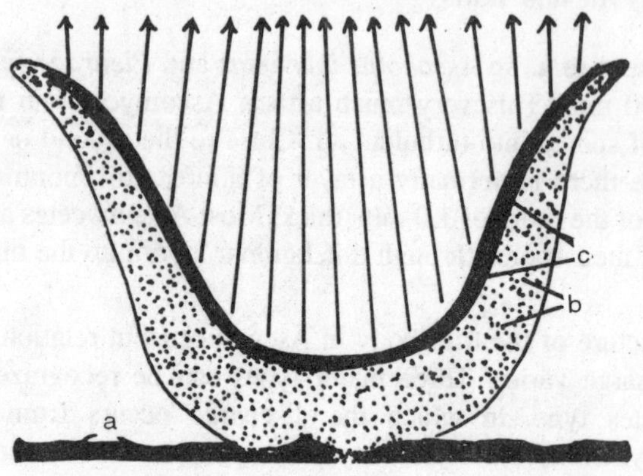

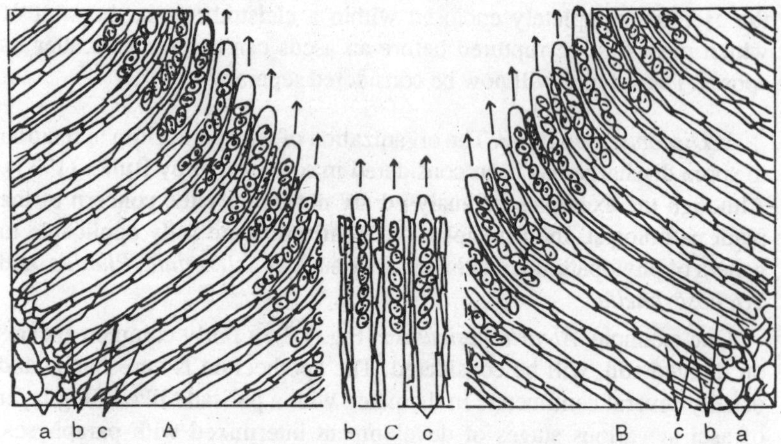

Fig. 4.2. Aleuria vesiculosa. Above: L.S. of small apothecium, *a*, substratum; *b*, hypothecium *c*, hymenium. The arrows show the direction of spore discharge. Below: parts of the hymenium from the sides (A and B) and the bottom (C) of the apothecium.

in humidity, as occurs when an apothecium is breathed upon, is enough to cause all these ripe asci to discharge simultaneously, sending into the air as visible cloud of hundreds of thousands of ascospores which drifts away like smoke. Buller has shown that this simultaneous bombardment sets the whole body of air just above the apothecium in motion, with the result that the spores are carried considerably farther than the distance to which a single ascus, discharging alone, could shoot its spores. Puffing may be of biological significance in the more efficient launching of the spores. Further, the cessation of discharge during periods of stillness tends to prevent the liberation of spores when the air is in a nonturbulent condition, that is, at such times when subsequent effective dispersal is less likely to occur.

2. *Pyrenomycetes Type*. Many Pyrenomycetes, especially in the genera *Nectria, Epichloe, Rosellinia, Mycosphaerella, Ophiobolus, Venturia, Endothia*, etc., are parasites of higher plants, but again it is more convenient to consider, in the first instance, spore liberation in a saprophytic type, namely *Sordaria fimicola*. The structure of this common coprophilous species is illustrated in Fig. 4.3. The flask-like perithecium

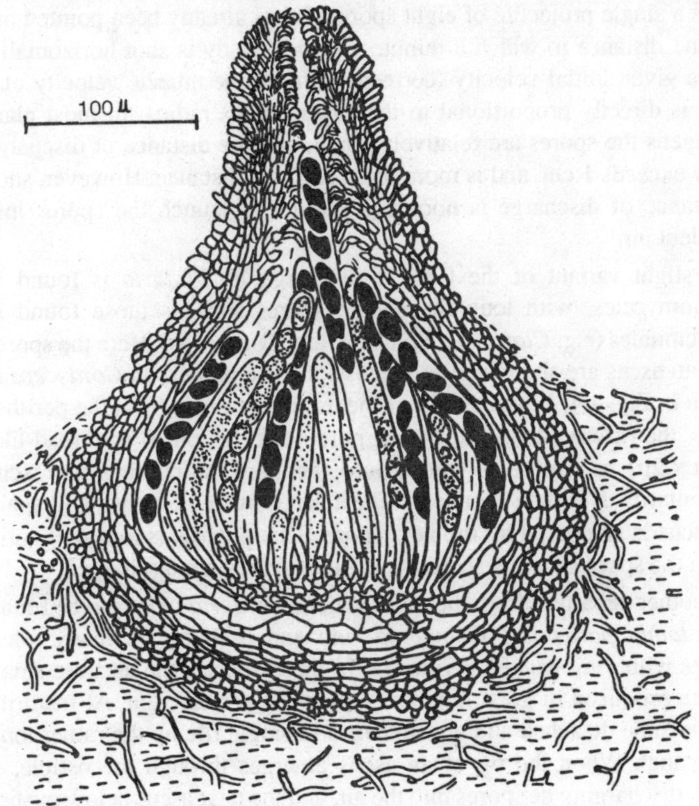

Fig. 4.3. Sordaria fimicola. L.S. perithecium growing on nutrient agar.

is filled with asci at various stages of development, and any free space between them is occupied by mucilage. In Pyrenomycetes generally no gas phase is present within an active perithecium. While remaining attached at its base, a ripe ascus elongates up the neck canal, which is lined by downward-projecting periphyses. When its tip protrudes slightly through the ostiole, the ascus bursts, shooting its spores to a distance of 1 to 10 cm. Another ascus then elongates, and so on, the discharge of asci occurring one at a time in orderly succession. There is obviously no opportunity for "puffing" to take place. Empty asci retract into the perithecium, and soon liquify and disappear. In *S. fimicola* the distance of discharge is relatively great, a feature of many coprophilous fungi, mainly because of the large size of the spores and because of the fact that they tend to stick together. The whole contents of the ascus often forms a single projectile of eight spores. It has already been pointed out that the distance to which a minute spherical body is shot horizontally, with a given initial velocity (corresponding to the muzzle velocity of a gun), is directly proportional to the square of its radius. In most plant pathogens the spores are relatively small, and the distance of discharge rarely exceeds 1 cm. and is more often half this distance. However, such a distance of discharge is normally enough to launch the spores into turbulent air.

A slight variant of the type of discharge in *Sordaria* is found in Pyrenomycetes, with long thread-like spores such as those found in Clavicipitales (e.g. *Claviceps, Epichloe*, and *Cordyceps*). Here the spores from an ascus are shot away in succession. The process in *Cordyceps* is shown in Fig. 4.4. When the tip of the ascus protrudes from the perithecium, the ascus dehisces by a pore, but at once a thread-like (300μ × 2μ) ascospore is forced into the pore, stoppering the ascus momentarily. It is then shot away like a dart from a blowpipe, and again the ascus is stoppered by the next spore. This process is repeated until, within the space of a few seconds, all eight spores are discharged.

Another departure from the more general *Sordaria* type is to be found in *Endothia parasitica, Gnomonia rubi* and probably in many other species with long-necked perithecia. In these species the asci are small and are produced in great numbers within the perithecium. At maturity they become detached and are squeezed in single file up the rather long neck canal. When the tip of an ascus emerges through the ostiole, it bursts, discharging its spores into the air, and the next ascus below pushes out the empty envelope of the first one. This process can be seen in operation in *Gnomonia rubi* (Fig. 4.5), where the neck of the perithecium

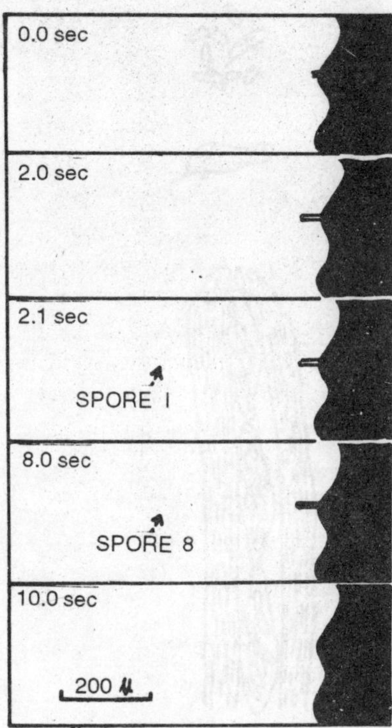

Fig. 4.4. *Cordyceps militaris.* Profile of two projecting perithecia as seen when a stroma is laid on its side and viewed under low power of microscope. At 0.0 sec. the tip of an ascus is beginning to project from its ostiole. At 2.0 sec. it has reached its maximum extension. At 2.1 sec. the first spore flashes into view, then rapidly falls out of sight. At 8.0 sec. the last spore has been discharged. At 10.0 sec. the empty ascus has retracted into the perithecium.

is sufficiently transparent for microscopic observation of the column of ascending asci. This type of perithecial behavior clearly allows very rapid spore liberation under suitable conditions, and less than a second may elapse between the discharge of successive asci. Such a speed of action would be impossible in *Sordaria*, where the interval between the bursting of successive asci is usually to be reckoned in minutes.

From this type of spore liberation it is but a short step to the condition in *Ceratocystic (Ophiostoma)*, where the asci not only become detached on ripening, but the ascus wall undergoes liquefaction. The naked asci, as discrete eight-spored droplets, still pass in single file up the perithecium

Fig. 4.5. *Gnomonia rubi.* Neck of the perithecium in optical section showing passage of asci along neck canal. Above the ostiole the spores and empty membrane of a discharged ascus are shown.

neck, but as they emerge through the ostiole, no discharge occurs, and they simply run together into a larger drop, poised at the end of the long, black neck. This step in evolution appears to be correlated with the transition from wind to insect dispersal.

3. *Erysiphales Type*. In the powdery mildews (Erysiphales) the cleistothecium is essentially a hibernating structure with a continuous hyphal wall around the ascus or asci within. In the spring, when conditions are favorable, the expanding asci must first rupture this wall before emerging to discharge their spores. This process is illustrated in Fig. 4.6 for *Sphaerotheca mors-uvae*.

In some powdery mildews the ascus as well as the ascospores may be violently discharged. This has been described, for example, in *Podosphaera leucotricha*. The expanding ascus ruptures the cleistothec-

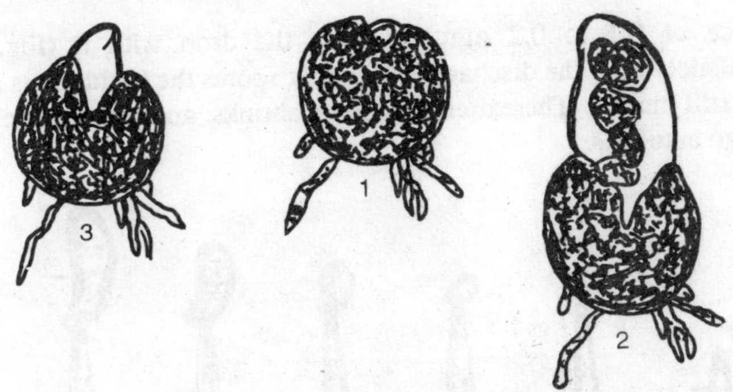

Fig. 4.6. *Sphaerotheca mors-uvae*. (1) swelling ascus is just bursting through cleistothecium wall; (2) fully swollen ascus is about to discharge its spores; (3) an instant later.

ium wall in an irregular manner. However, this wall itself is elastic. As the enlarging ascus protrudes, it tends to slip out of the stretched cleistothecium shell. The jaws of this shell suddenly spring together, ejecting the whole ascus several centimeters into the air. The ascus itself then bursts, scattering the spores.

B. Basidiomycetes

In Basidiomycetes violent spore liberation is characteristic of Hymenomycetes, Tremellales, Uredinales, and certain Ustilaginales. In Gasteromycetes, in which the whole hymenomycete equipment of spore liberation appears to have been lost, the basidium is no longer a spore gun, and consequently, novel methods of spore liberation have been developed along diverse lines. However, Gasteromycetes, being essentially saprophytic, are no concern of the plant pathologist.

The basidium of the Hymenomycetes consists of a single cell bearing apical sterigmata, usually four, with a basidiospore at the end of each. The basidiospore is poised asymmetrically on its sterigma, and when the basidium is ripe, its four spores are discharged in succession. On the spore, very near its junction with the sterigma, is a minute projection—the hilum. It is this hilum that gives the highly characteristic appearance to discharged basidiospores, allowing them to be spotted readily on slides exposed by the aerobiologist. Just before a basidiospore is to be shot away, a drop of liquid makes its appearance at the hilum, and grows to a certain definite size; then the spore is discharged, usually to a distance of 0.1 to 0.2 mm., carrying the drop with it (Fig. 4.7). Immediately after the discharge of the four spores the basidium is apparently still turgid. Thereafter it slowly shrinks and finally seems to undergo autolysis.

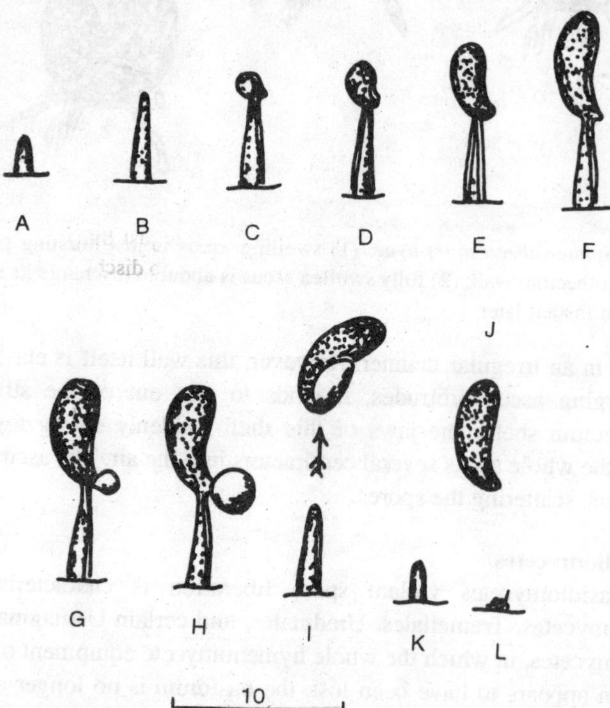

Fig. 4.7. *Calocera cornea.* A-L, stages in the development of the basidiospore at the end of the sterigma and subsequent spore discharge.

Although studied fairly extensively, the mechanism of basidiospore discharge remains a mystery. It has been suggested (and for this there is good observational evidence in the rust *Gymnosporangium juniperivirginianae*) that at maturity a cross-wall separates spore and sterigma. Discharge might thus be due to the sudden rounding-off at this region of the junction between a turgid spore and a turgid basidium.

If this is so, basidiospore liberation is essentially like conidium discharge in *Conidiobolus* or *Sclerospora*. However, because the sterigma is so narrow at its junction with the spore and because in the fully mature condition observation can only be made in air (since mounting in water at once breaks the connection) it is impossible to be sure whether a cross-wall exists at the moment of discharge. Again, if the discharge is due to rounding-off at the interface between spore and sterigma, the beautiful asymmetry of its position on the sterigma would seem to have no significance, nor would the process of drop secretion at the hilum, which so regularly heralds discharge.

Another theory is that the basidiospore is discharged from its sterigma by a water-squirting mechanism. According to this view the turgid basidium bursts at the end of the sterigma, discharging a liquid jet which carries the spore with it. Against this view is the fact that immediately after a spore has disappeared, the vacant sterigma appears to be closed at its apex, and no exudation of fluid normally takes place from it. Indeed if this type of discharge is to occur, the sterigma must become sealed immediately so that the turgidity of the basidium may be maintained for the discharge of the remaining spores. As a slight modification of this view, it has been suggested that the drop is exuded not from the spore itself but at the junction of the spore and the sterigma. It is supposed that, having grown to a certain size, the drop is shot away, carrying the spore with it. This view is supported to some extent by Müller's study of spore discharge in the mirror yeast *Sporobolomyces*, in which the aerial conidium is discharged in a manner that appears to be identical with basidiospore discharge in Hymenomycetes. Müller, (1954) in a cinematographic study showed that, although normally the spore and its exuded drop disappear simultaneously from the end of the sterigma, occasionally the drop suddenly disappears (presumably having been discharged) leaving the spore behind on the sterigma. However, in this connection it must be remembered that the careful observations of Buller (1933) all go to show that the drop exudes from the spore itself and not from the junction of spore and sterigma.

It has also been suggested that the surface energy of the exuded drop might be mobilized in some way to effect spore discharge. This view has the merit of bringing the exuded drop and the asymmtrical poising of the spore into the picture, but unfortunately there is no real evidence to support it, nor is it at all clear how this energy could be mobilized. It has been calculated, however, that there is enough surface energy available, if it could be utilized, to discharge the spore to the observed distance.

The essential features governing fruit-body construction in Hymenomycetes are the very short distance of basidiospore discharge and the fact that, since basidiospores are sticky, they cannot normally be dislodged by wind once they have settled on a surface. Upward-facing hymenial surfaces are very rarely encountered in Hymenomycetes. Spores shot vertically to a distance of only 0.1 to 0.2 mm. would not usually reach the turbulent air above the laminar layer in contact with the hymenium, and would fall almost at once on the hymenial surface and become permanently stuck. In fact the hymenium in the sporophores of Hymenomycetes is mostly vertical, although it is sometimes horizontal and downward-facing, or it may occupy an intermediate, but still downward-facing, position. It should, perhaps, be remarked here that the position assumed by the hymenium in most toadstools and bracket fungi also tends to protect them from rain. This is of importance, since water on the surface disorganizes the hymenium, at least temporarily, in striking contrast to that of Discomycetes, which are not injured by the temporary presence of free water.

Since, for the plant pathologist (especially if he is a forester) the bracket polypores are probably of more significance than agarics, spore liberation in the larger Basidiomycetes will be considered in some detail in a polypore and more briefly in agarics, although the general principles in both are essentially similar.

As a first example, *Ganoderma applanatum* will be taken. This fungus, with an almost world-wide distribution, is a wound pathogen of mature broad-leaved trees causing a heart rot. The mycelium occurs in the wood and the rigid sporophore forms a bracket or shelf broadly and firmly attached to the tree trunk. This woody fruit body is a perennial structure, the small specimen illustrated in Fig. 4.8 being 2 years old. Examples of 5 or 6 years of age are frequently to be found. The upper surface is extremely hard. On the underside the hymenial tubes grow down from the sterile cap tissue, and these are long (up to 3 cm. in the specimen figured) and extremely narrow (0.1 to 0.2 mm. diameter) (Fig. 4.9). During the growing season (May to September) they elongate, the meristematic growing region being at the orifice of each tube. Growth ceases in autumn and winter, and is resumed in late spring. The arrest of growth has structural implications, and the result is that if a sporophore is broken vertically, successive annual layers of growth are clearly visible. Growth of the tubes is controlled by gravity so that each individual tube develops in a perfectly vertical manner. This, combined with the rigidity of the fruit body and its firm attachment to a stout trunk, ensures that the tubes maintain their perfect verticality. This seems to be essential for effective spore liberation. Active hymenium lines each tube

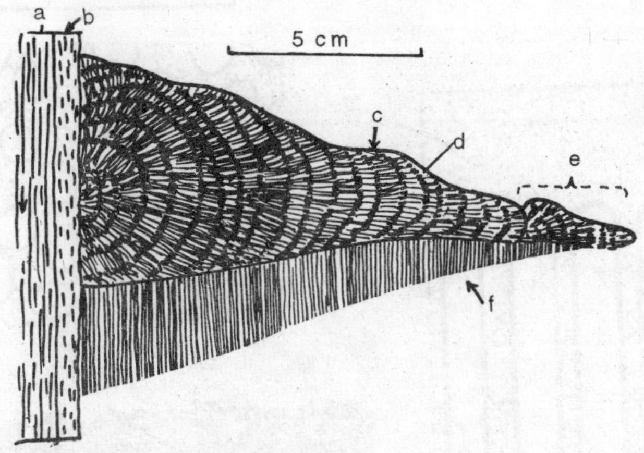

Fig. 4.8. *Ganoderma applanatum*. Vertical section of a small sporophore in its second year growing on ash. *a*, wood; *b*, bark; *c*, upper "crust" of fungus; *d*, zoned fibrous cap tissue formed in first year; *e*, additional cap tissue produced in second year; *f*, hymenial tubes (the dotted line indicates the boundary between the tubes formed in the first year (above) and in the second (below).

throughout most of its length, so far at least as the last two or three annual layers are concerned. In still older parts of the tubes the hymenium ceases to be functional, and the pores may become blocked with sterile hyphae.

The spores in this species are discharged horizontally to a distance of approximately 0.05 mm. They then fall under the influence of gravity down the narrow tube. If the tube were the merest fraction of a degree out of the vertical, clear fall would seem to be impossible and the spores would presumably become stranded on the hymenial surfaces lower down the tube. There is, however, just the possibility that the spores, which have been shown usually to carry a positive static electric charge, are maintained in midstream during their fall down the hymenial tube by static electric forces.

On emerging into the free air below the tubes, the spores are carried away by the wind. It is to be noted that by their very situation on the tree trunk, the sporophores are in a position to drop their spores directly into the turbulent air. As if to achieve this end, ground-inhabiting species such as bolets and most agarics are forced to raise their spore-producing surfaces above the substratum on stripes. As Gregory (1952) has pointed out, the Ascomycetes can shoot their spores through the surface laminar

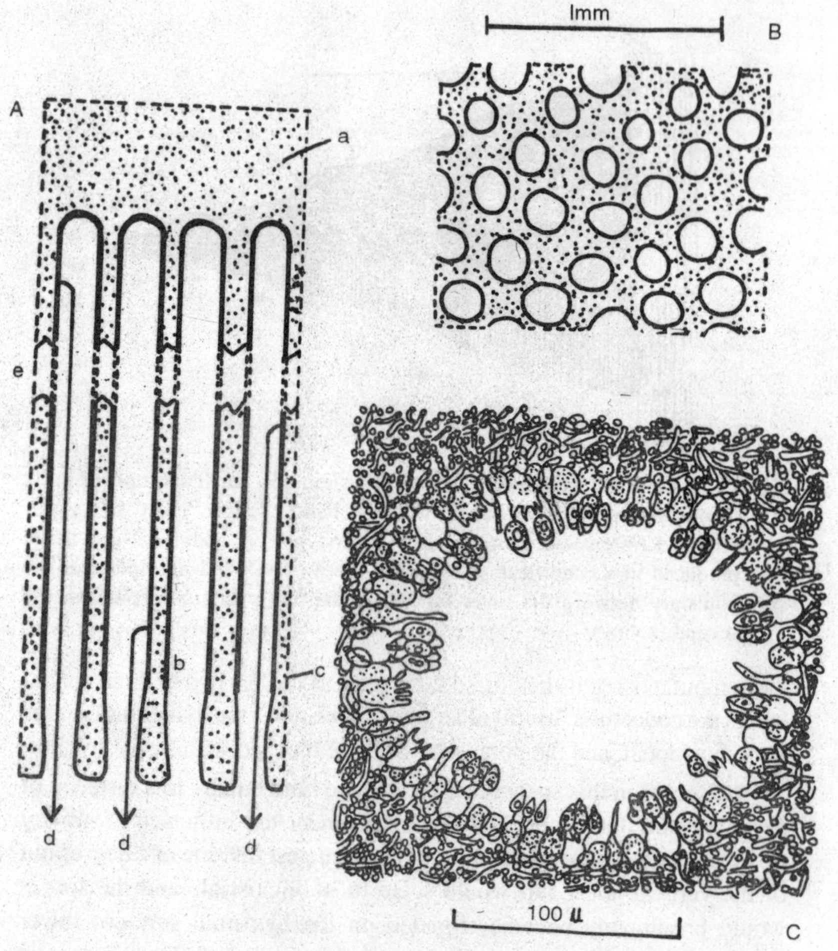

Fig. 4.9. *Ganoderma applanatum.* A. Diagram of longitudinal section of hymenial tubes. *a*, cap tissue; *b*, hymenial tube; *c*, active hymenium shown by thick black line; *d*, trajectory of discharged spore; *e*, gap in the diagram corresponding to about 200 mm. in the actual specimen or 574 mm. at the scale of this figure. B. Transverse section in region of hymenial tubes. The thick black line arround each pore is the hymenium. C. Drawing of a single hymenial tube seen in transverse section.

layer of air into the turbulent layer above, but the basidium is not a sufficiently powerful spore gun to achieve this, so that in Basidiomycetes the hymenial surfaces tend to be so placed that the discharged spores drop into turbulent air. This concept is illustrated diagrammatically in Fig. 4.10.

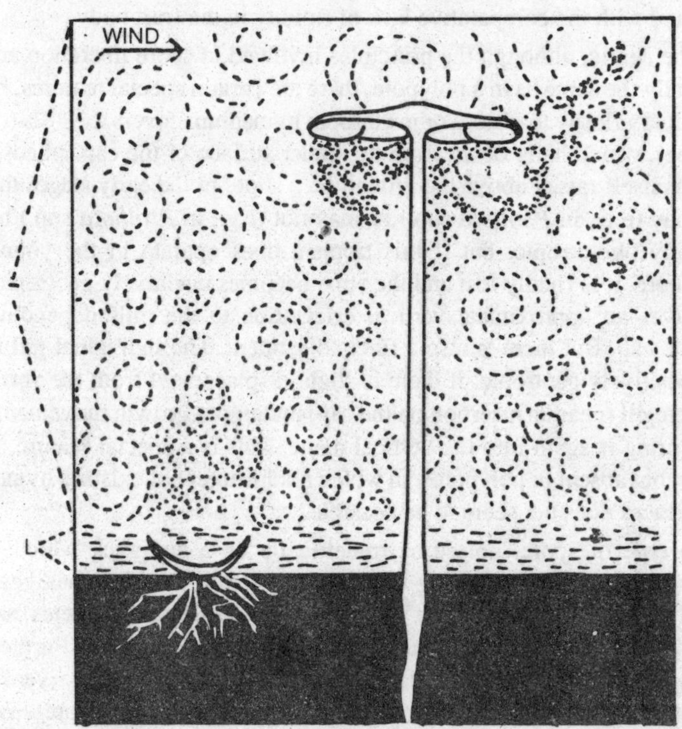

Fig. 4.10. A cup-fungus (*Peziza* sp.) and an agaric (*Collybia radicata*). The ground is shown black. Above this is a thin layer of laminar air (L) and on top of this turbulent air (T). The *Peziza* has just discharged a puff of spores through the laminar air into the turbulent region. From the pileus of the *Collybia* spores are dropping into the turbulent air.

In *Ganoderma applanatum* spore liberation is on a gigantic scale, and Buller has estimated that a large sporophore may liberate millions of spores a minute, maintaining this rate more or less for the whole 6 months of its annual spore-fall period.

Ganoderma applanatum is an extreme example which serves to emphasize the principles of basidiospore escape from a sporophore. Apart from species of *Ganoderma* and *Fomes*, bracket polypores have annual sporophores, and usually the pores are wider and shorter than those of *Ganoderma*. Thus in *Polyporus squamosus*, a common species causing a heart rot of broad-leaved trees (especially elm), the pores are 1 to 2 mm. wide and only about 1 cm. long. These features may be correlated with the comparative lack of rigidity in the fruit body.

In the agaric, although the principles involved in spore liberation are essentially the same as in a polypore, there are certain special features. In the ordinary fleshy toadstool or mushroom hymenium covers the gills (or lamellae), which hang down from the under surface of the cap (pileus), which is itself raised above the ground on a stipe. In the early stages the stipe may (e.g., in *Pholiota* spp.) or may not (e.g., in *Psalliota* spp.) be positively phototropic, but if this tropism does operate in the young sporophore, it is finally lost and the stipe becomes negatively geotropic. This gives an approximate vertical orientation to the gills depending from the cap. But there is also a finel adjustment. The individual gill is itself positively geotropic. If there is slight displacement from the vertical, each gill (near its insertion on the cap) undergoes growth movements which bring it again into the vertical plane. This is a special feature of agarics, but absent in polypores, in which readjustment of existing hymenial surfaces does not seem to be possible.

The type of spore, poised asymmetrically on a sterigma, which is violently discharged following drop secretion from a hilum has been termed a "ballistospore." The basidiospores of Hymenomycetes are ballistospores; those of Gasteromycetes and of *Ustilago* are not. Further, certain ballistospores, especially those of the shadow yeasts (Sporobolomycetaceae) cannot easily be regarded as basidiospores. We must now give some consideration to the ballistospores of rusts (Uredinales), which are clearly basidiospores, and to those of *Tilletia caries*, which can be classified as such only by a rather tortuous argument.

In rusts, the teliospore, usually after a winter's rest, germinates to produce a curved transversely septate basidium (promycelium) bearing four basidiospores (sporidia) on its convex side. This organization of the telium in relation to the discharge of basidiospores in rusts has been illustrated by Buller (1931) in *Puccinia malvacearum* (Fig. 4.11). It is clear that the curvature of the basidium and the position in which basidiospores arise is of importance in relation to the free escape of the discharged spores.

In smuts the occurrence of ballistospores has been demonstrated, particularly in *Tilletia caries*, but it is clear that in *Ustilago* spp. violent discharge does not take place. Buller (1933) has shown that when the brand spore of *T. caries* germinates on a moist surface a short promyce-

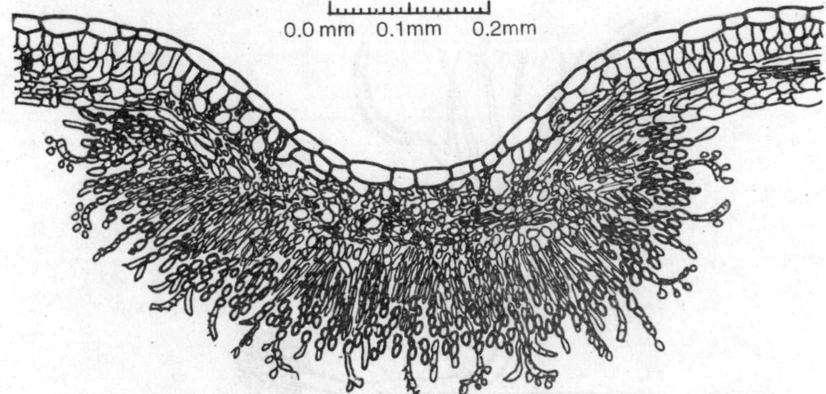

Fig. 4.11. *Puccinia malvacearum.* A sorus of germinating teliospores on the under side of a leaf of hollyhock (*Althaea rosea*).

lium is formed. This becomes septate and soon grows away from the surface, probably because of negative hydrotropism. From the end cell a tuft of six to twelve needle-shaped spores (primary conidia) is formed. These have been regarded by most mycologists as the basidiospores. While still attached, or after passive liberation, they conjugate, forming *H*-shaped pairs. From one member of each pair a short sterigma is produced bearing an asymmetrically poised, sickle-shaped spore (secondary conidium) which is a typical ballistospore being violently discharged following drop excretion at the hilum (Fig. 4.12). If an *H*-shaped pair of conjugated primary conidia is planted on a nutrient medium, such as malt agar, a fine, branched mycelium may be formed from which, in due course, a number of sickle-shaped ballistospores may arise on aerial sterigmata. In the field the ballistospore may germinate in contact with a host plant and cause infection, but on nutrient agar, and possibly under certain conditions in nature, a saprophytic mycelium is formed from which further ballistospores are produced. It is of interest to note that in *Tilletia caries* the record discharge distance for a ballistospore has been observed, namely, 1.0 mm.

C. Other Types of Discharge

We must now consider a number of examples in which the essential mechanism of discharge involves the sudden rounding-off of turgid spores. Probably the most important example from the point of view of plant pathology is the aecium of rusts (Uredinales) from which the aeciospores are commonly shot to a distance of 5 to 15 mm. Nevertheless

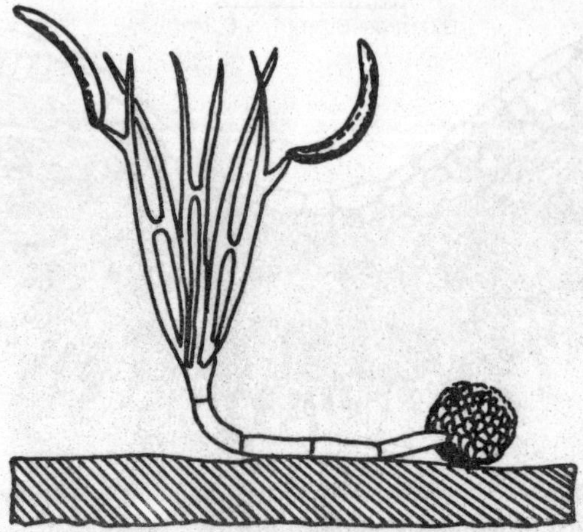

Fig. 4.12. *Tilletia caries*. Germinating chlamydospore producing a promycelium bearing a crown of six filamentous primary conidia which, following pair-wise conjugation, are producing secondary conidia which are ballistospores.

in some species no violent liberation occurs. Although the force of propulsion is clearly provided by the sudden change in form of turgid spores, the details of the process are not absolutely clear. The mature aecit m is usually a cup-like structure with a firm wall (pseudoperidium) of thickened cells. At the base of the cup is a close-set palisade of basal cells, each producing an ever-growing file in which ascospores and intercalary cells alternate. Actually the basal cell cuts off a single terminal cell at a time; this divides to form a larger cell above, which becomes the spore, and a smaller below, which is the intercalary cell. Often this cell is delimited not by a wall parallel with the flat top of the basal cell, but by a curved wall at an angle to this plane. The result is that the intercalary cell arises in a corner-wise position—rather as a companion cell is carved off from a sieve tube. The intercalary cell remains thin-walled, while the wall of the associated aeciospore thickens. Most workers have regarded this cell as an ephemeral structure, which soon breaks down, but it has been suggested that it remains turgid until the maturation of its companion spore.

As the asciospore develops, its wall differentiates into three layers, and at the same time the future germ pores are organized. In some rusts where a pore is to be formed, the wall is much thicker locally and probably different chemically, producing a minute spherical plug, which becomes more or less free from the rest of the wall. In the spore deposit from a discharging aecium, these plugs can be seen either still adhering to the spores or lying free (Fig. 4.13). When the plug is eventually

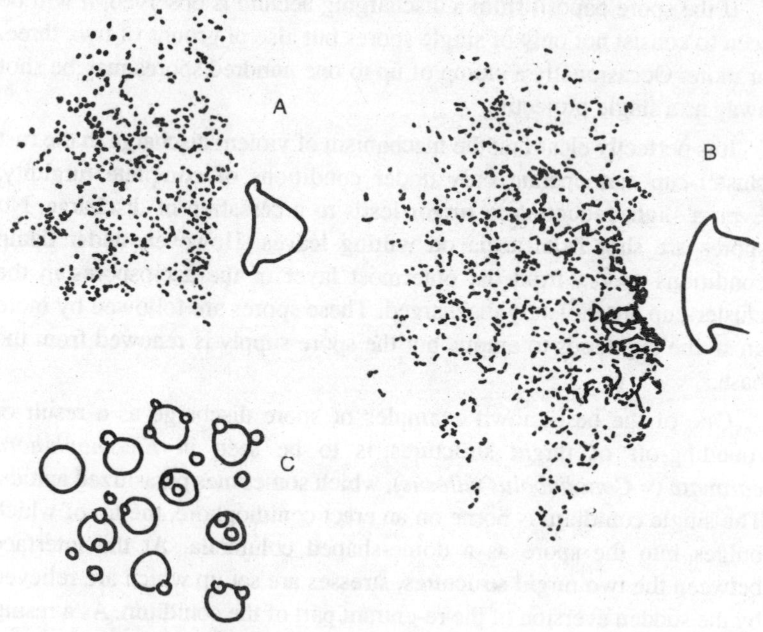

Fig. 4.13. *Gymnosporangium myricatum* A. and B. Spore prints formed after dissecting an aecium and laying it down on a glass slide in a damp chamber. Prepared from a photograph, C. Outline of spores showing plugs that still remained attached to spores afer their flight and also free plugs.

displaced, it leaves a very thin region in the spore membrane through which a germ tube may emerge if the spore eventually germinates.

It has been suggested that these spore plugs may play an essential part in asciospore discharge—a part which can be illustrated by a simple model. "A tennis ball compressed over a marble on a table will be thrown farther upward when the confining pressure is suddenly removed than it will be if compressed against the table alone with the same force, and then released." The suggestion is that in the closely packed aecium the spore is indented by the pore plugs, but in the end the spore suddenly rounds off and is discharged. The pore plug acts as the marble in the model. The major difficulty of this theory is that in some rusts the aeciospores are actively discharged in spite of the absence of pore plugs.

It may be, however, that the turgid, intercalary cells persisting among the mature spores act in the same way.

If the spore deposit from a discharging aecium is observed, it will be seen to consist not only of single spores but also of groups of two, three, or more. Occasionally a clump of up to one hundred spores may be shot away as a single projectile.

It is perfectly clear that the mechanism of violent discharge in the rust cluster-cup can operate only under conditions of complete turgidity. Even a slight reduction in turgor leads to a cessation of discharge. No spores are shot from aecia on wilting leaves. However, under damp conditions spores from the outermost layer of the aeciospores in the cluster-cup are violently discharged. These spores are followed by more an so the cup tends to empty, but the spore supply is renewed from the base.

One of the best-known examples of spore discharge as a result of rounding-off of turgid structures is to be seen in *Entomophthora coronata* (= *Conidiobolus villosus*), which sometimes parasitized aphids. The single conidium is borne on an erect conidiophore, the tip of which bulges into the spore as a dome-shaped columella. At the interface between the two turgid structures, stresses are set up which are relieved by the sudden eversion of the re-entrant part of the conidium. As a result, the spore springs into the air to a distance of a few centimeters.

The same type of discharge has been reported in *Sclerospora philippinesis*, causing downy mildew of maize. In this fungus, however, the tip of the conidiophore branch does not project into the conidium, but there is a flat surface of contact between the two. It seems that it is by rounding-off in this region that spore discharge occurs. (Fig. 4.14). In *S. philippinensis* the spore is shot to only a very short distance—about 1 mm. This is probably due to the relatively small area of contact between the reacting structures.

It has been reported that the rounding-off mechanism of violent discharge operates in some powdery mildews (Erysiphales). Hammarlund (1925) fitted capillary tubes over single conidiophores and watched spore liberation microscopically. He maintained that the mature conidium is shot to a distance of several spore lengths due to sudden rounding-off at the junction with the next spore in the chain. This observation, however, lacks confirmation, possibly because no later worker has used such an elegant techniques as that employed by Hammarlund.

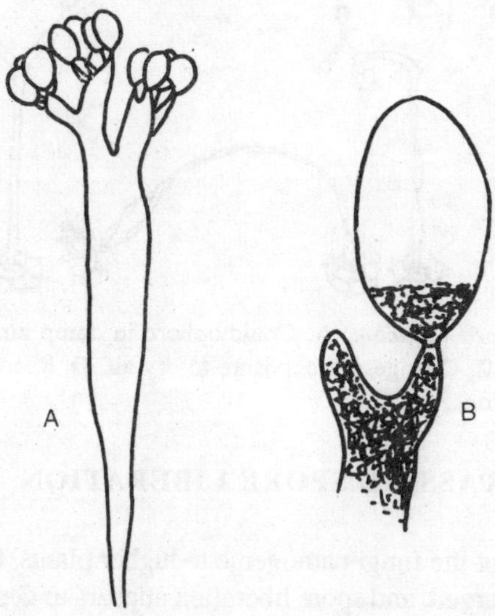

Fig. 4.14. *Sclerospora philippinensis*. A. Conidiophore bearing eleven conidia. B. Part of conidiophore more highly magnified; from the tip on the left the conidium has been discharged, that on the right being still attached.

Although in *Sclerospora*, as we have seen, spores are discharged by a rounding-off mechanism, generally in downy mildews (Peronosporaceae) violent discharge is not very common. However, in a few other species it does occur, but by a mechanism quite unlike that found in maize mildew. Pinckard (1942) has described a case of this kind. In *Peronospora tabacina* (blue mold of tobacco) as the branched conidiophore dries, violent twisting movements occur which may flick the finelly attached spores into the air (Fig. 4.15). It might be supposed that even without these movements, the ripe spores would readily be detached by wind, and thus the biological significance of this hyposcopic discharge would be in doubt. However, Waggoner and Taylor (1958), as a result of spore-trapping carried out over tobacco seed-beds heavily infected with blue mold, found that spores were trapped mainly in the morning, when hyposcopic twirling of the drying conidiophores would be expected to occur. The absence of spores in the air during midday hours, when strong hyposcopic movements would be at a minimum, was taken to imply that forcible discharge is a necessity for spore liberation.

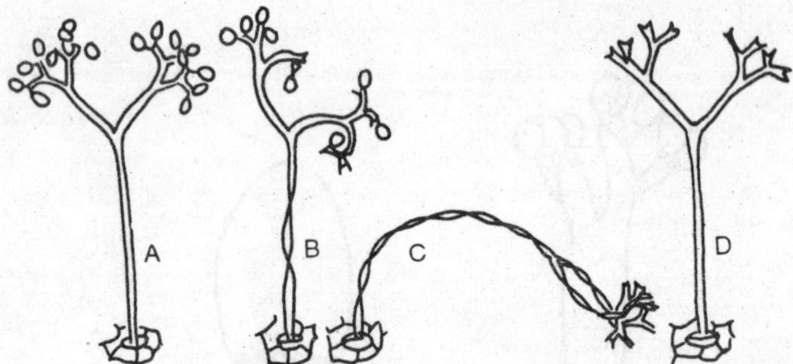

Fig. 4.15. *Peronospora tabacina*. A. Conidiophore in damp air with attached conidia. B and C. Changes on exposure to dry air. D. Recovery on return to damp conditions.

II. PASSIVE SPORE LIBERATION

In a large number of the fungi pathogenic to higher plants, the spores are not violently discharged, and spore liberation appears to depend on wind, rain splash, or the activity of insects.

In considering the classification of conidial fungi (Hyphomycetes or Moniliales) Mason (1937) laid emphasis on the biological importance of the difference between dry-spore types (Xerosporae) and slime-spore forms (Gloiosporae). In the former the conidial could be liberated by wind, but in the latter wind could not bring about liberation, and water or insects would seem to be indicated as agents in dispersal. However, from the point of view of dispersal the difference between Mason's two types is not always clear-cut. Thus is some dry-spore fungi even the strongest winds fail to dislodge spores from their conidiophores, while some slime-spore species form their conidia in spore drops which eventually evaporate, leaving dry spores which can be blown off the conidiophore either singly or in groups. A common example is *Trichoderma viride*, a saprophyte in soil or on wood.

In a number of dry-spore species which rely on wind for spore liberation, the ripe dispersive units (propagules) are borne some distance above the surface of the host on erect structures. This is so in such diverse fungi as downy mildews (Peronsporaceae), gray mold (*Botrytis cinerea*), powdery mildews (Erysiphales), *Omphalia flavida* (gemmifers with gemmae), *Cladosporium, fulvum*, and many others. The advantage seems to be that the propagules are exposed directly in turbulent air above the nonturbulent layer, perhaps 50 to 100μ thick, and which so often exists as a skin over the host surface. However, there are many dry-spore pathogenic fungi in which the spores are not raised to an appreciable extent above the surface of the diseased tissue. In this connection special mention must be made of most uredospores of rusts and brand spores of smuts. In these fungi the spores are, without doubt, freely liberated into the air in spite of the absence of violent discharge or of conidiophores raising the spores into a favorable position for liberation. It seems that the spores of such fungi, provided that they are sufficiently loose and powdery when mature, can be liberated freely when the diseased leaf, stem, or inflorescence on which they occur vibrates in the wind. This is the ancient principle of the sling shot. Indeed, it would seem quite likely that long conidiophores only acquire an importance when fungi are very close to ground level or when the part of the host attacked is rigid. These remarks only serve to underline our basic ignorance of this question. Further, we do not know in many cases whether spores become detached more readily under dry or under damp conditions.

In slime-spore fungi there is often doubt about the manner of their dispersal, although the agents concerned are likely to be insects, or water, or both. Among slime-spore fungi well-known to the phytopathologist are *Fusarium* spp., *Collectotrichum* spp., the conidial stage of *Nectria* spp., most members of Sphaeropsidales, the pycnial stage of rusts, the *Sphacelia* stage of ergot, and *Graphium* spp. In the last three examples insect dispersal of the minute sticky spores has been established, but in the others rain splash is probably the main factor involved in the take-off of the spores from the parent tissue. A well-known example of dispersal of rain splash is *Colletotrichum lindemuthianum*, the cause of anthracnose of *Phaseolus* spp. (dwarf and runner beans). The fungus attacks the aerial parts of the host, producing dark, sunken lesions bearing pink acervuli. The slime spores from these are scattered by splashing rain onto healthy foliage and pods.

At this points brief mention should be made of the take-off from the host of bacteria pathogenic to higher plants. In bacteria, apart from certain Actinomycetes, there is no parallel with dry-spore conidial fungi. However, in their dispersal there is a close parallel with slime-spore fungi. In a number of bacterial plant diseases, at some stage slimy masses of bacterial cells occur on the affected host, and the dispersal of these is due either to insects (as *Bacterium amylovorum*, causing fireblight of pear and apple) or to rain splash. As example of the latter is *Xanthomonas citri*, the cause of citrus canker. The bacteria exude from scabby spots on leaves, young twigs, and fruits, and rain splash carries infection from diseased to healthy tissue.

The problem of the basic mechanics of splash dispersal has recently been investigated by Gregory and his co-workers (1959). The technique used was to allow water drops of definite size to fall onto a watery spore suspension exposed as a film of known thickness on a glass slide. The size and scatter of the reflected droplets was studied and they were examined to determine whether they carried spores. As a test organism *Fusarium solani*, a slime-spore species, was generally used. It was found that a drop 5 mm. in diameter falling from a height of 7.4 m. onto a spore-containing film 0.1 mm. thick produced over 5000 reflected droplets of which more than 2000 carried spores. The droplets ranged in size from 5μ to 2400μ. On the average the distance which they were scattered horizontally was 10–20 cm.

It is to be noted that the larger reflected droplets fall back within a small fraction of a second onto the substratum. The smaller ones, however, may remain longer in the air and be rapidly reduced further in size by evaporation. Thus, as a result of rain splash, slime spores may become suspended in turbulent air and be dispersed in the same manner as dry spores. Although the lists of species recorded by aerobiologists invariably show a great preponderance of dry-spore fungi, there is always a slime-spore element and rain splash may be the principal factor involved in the contribution of this to the air spora.

A study, involving high-speed photography, was also made of splash liberation of conidia from a twig bearing abundant conidial stromata of *Nectria cinnabarina*. Large drops (5.0 mm. diameter) falling from a height into the twig each broke into thousands of droplets all of which carried spores.

Much further quantitative study, particularly in the field, is needed in connection with the problem of splash dispersal, but the work of Gregory and his colleagues provides an inspiring model for future research.

III. METEOROLOGICAL CONDITIONS IN RELATION TO SPORE LIBERATION

Liberation of fungal spores may be conditioned external factors—especially humidity, rain, temperature, wind, and light.

A. Humidity and Rain

These two factors, although usually correlated, are sometimes separable in their effects. As we have seen, rain may have a special effect in splash liberation of certain spores, but, further, some fungi require actual wetting if spore discharge is to take place. This is particularly true in Pyrenomycetes. Thus, for example, in *Nectria galligena*, in *Venturia inaequalis* and in *Endothia parasitica* the discharge of ascospores is closely correlated with rainfall (Fig. 4.16).

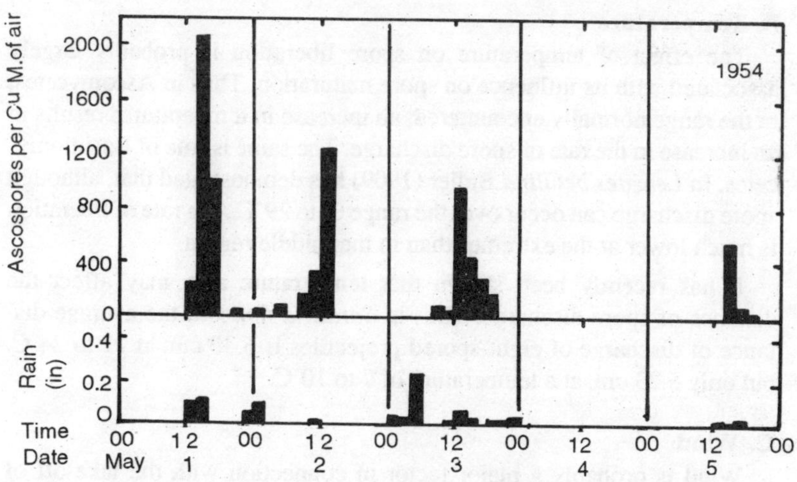

Fig. 4.16. Relation of rainfall and concentration of *Venturia inaequalis* ascospores per cubit meter of air every 2 hours during May, 1954.

In some species (e.g., *Hypoxylon pruninatum*, it has been shown that a saturated atmosphere is not sufficient in itself to initiate spore discharge but that the perithecial stroma must first be wetted before discharge will begin. Although this relationship in Pyrenomycetes between spore discharge and rainfall is the normal one, there are exceptions. Thus *Daldinia concentrica* can discharge its spores under the driest of conditions, obtaining the water necessary for continued discharge from a water reserve in the stromatal tissue, and *Epichloe typhina* can also discharge its spores even when the stroma is unwetted and the surrounding air is dry, relying on the water brought to it in the transpiration stream of the host.

High atmospheric humidity and rain are important for continued spore discharge not only in Ascomycetes but also for liberation of basidiospores and for the escape of aeciospores in rust. However, in a number of dry-spore fungi, low humidity may be important for the take-off. In unpublished work in my own department it has recently been established that the sporangioles of the mold *Thamnidium elegans* and the conidia of *Trichothecium roseum* become detached by wind much more readily if it is dry than if it is damp, and the same may be true of other dry-spore fungi.

B. Temperature

The effect of temperature on spore liberation is probably largely associated with its influence on spore maturation. Thus in Ascomycetes, in the range normally encountered, an increase in temperatures results in an increase in the rate of spore discharge. The same is true of Basidiomycetes. In *Lenzites betulina* Buller (1909) has demonstrated that, although spore discharge can occur over the range 0° to 29°C., the rate of liberation is much lower at the extremes than in the middle region.

It has recently been shown that temperature also may affect the distance of spore discharge. Thus in *Sordaria fimicola* the average distance of discharge of eight-spored projectiles is 6.30 cm. at 21 to 24°C. but only 5.23 cm. at a temperature of 7 to 10°C.

C. Wind

Wind is probably a major factor in connection with the take-off of many parasitic fungi although quantitative information is sadly lacking. Mean-wind speed increases in a regular manner with height, and this may be expressed by the formula.

$$v_2 = v_1 z^{\alpha}$$

where

v_1 = wind velocity at unit (1 meter) distance above the ground

v_2 = wind velocity z units above the ground

and

α = an exponent which varies to some extent but near the ground is usually in the range of 0.2 to 0.3

This is, of course, a statistical "law", true on the average, but not necessarily giving a correct picture at any given moment. Theoretically at ground level the wind velocity is zero. Even under gale conditions a layer, perhaps only a few molecules thick, is still, or, if it is in motion, is nonturbulent. This "laminar layer" during the day is often a millimeter or less in thickness, but during the night with the disappearance of thermal turbulence, it may be reckoned in meters. However, although this laminar layer is normally present, it may locally and temporarily be invaded by violent eddies which may pick up dry spores and whirl them upward.

In connection with wind as a spore-liberating agent, the position of the spore-producing surfaces on diseased plants and the disposal of the spores on the conidiophores may be of great importance. It is to be hoped that wind-tunnel experiments will in the future add much to knowledge of the influence of wind on the escape of spores.

D. Light

Light is a minor factor in connection with the take-off of the spores of pathogens. Like temperature, it probably exerts its influence, if any, in conditioning spore maturation. It will be discussed later in connection with periodicity of discharge.

IV. PERIODICITY OF SPORE LIBERATION

From the point of view of the spread of plant diseases, the time of day when the spores of a pathogen are set free may be of importance. If the spores are liberated when atmospheric turbulence is high, their wide dispersal is favored, but this is likely to be an advantage only if they are of the type that can survive the relatively dry conditions often associated with turbulence. For delicate, short-lived spores it may be vital to the spread of the pathogen that liberation occur at a time of day when the air is damp and when potential infection drops, as the result of dew or rain, are likely to persist on the host leaves. Thus any daily cycle of spore liberation must be carefully studied by the plant pathologist.

Broadly speaking, this question of periodicity has been investigated in two different ways. First, the actual liberation of spores has been studied, often in the laboratory, and second, the daily cycle of spore content in the air in the neighborhood of infected plants has been followed.

In considering any daily cycle of spore liberation in a particular case it must be borne in mind that each of the four major factors influencing the take-off of spores, namely temperature, humidity, light, and wind, tends to exhibit a daily cycle of behavior. Light, temperature, and wind velocity tend to be maximal in daytime, while humidity tends to be high at night associated with lower temperature and decreased atmospheric turbulence. Again, it is, perhaps, important to realize that periodic appearance of spores in a spore trap does not necessarily mean that they have been set free in a periodic manner. Let us consider, say, the apothecium of a Discomycete on the ground discharging spores at a uniform rate. During the night in fair weather there may be a thickish layer of still or laminar air close to the ground, and spores discharged into this would soon fall to the ground and would not be caught in a spore trap operating at the standard height of 1 m. During the day, with the onset of turbulent conditions and the reduction of the laminar layer to a bare millimeter or so, spores would be brought into the region sampled by the trap. Thus a periodicity might be recorded quite unconnected with spore liberation. Probably, however, there is normally, a close correlation between libertine and trapping, provided the distance between the source and the trap is not too great.

Another general point which should be made is that periodic rhythm in spore liberation may be conditioned by periodic spore production or by the periodic operation of conditions which favor the escape of spores from a reservoir in the parent tissue.

In the powdery mildew of clover, caused by *Erysiphe polygoni*, it has been shown that each conidiophore normally produces and abstricts one conidium each day and that this rhythm depends on the natural alternation of light and darkness. Spore liberation tends to occur during the daytime and in greatest abundance about noon. Childs (1940) has also reported a diurnal cycle of spore maturation in a number of other powdery mildews, including species of *Sphaerotheca* and *Podosphaera*, and Yarwood (1957) has found a similar state of affairs in species of *Micosphaera* and *Uncinula*. However, in *Erysiphe graminis* on barley on diurnal cycle was observable.

In *Taphrina deformans* a periodicity has been found with spores discharged in greatest numbers in the evening. For the basidiomycete *Corticium filamentosa*, which forms its basidia on leaves of *Hevea*, nocturnal spore discharge has been demonstrated. Nocturnal discharge of ascospores occurs in *Daldinia concentrica*. This is an endogenous rhythm which continues for a number of days in continuous darkness at constant temperature. Under these conditions, however, the rhythm is finally lost, but is immediately restored on return to the normal periodic alternation of light and darkness.

In the saprophytic pyrenomycete, *Sordaria fimicola*, the periodicity of spore discharge has also been rather fully studied, and it may be that the picture obtained will prove to be generally applicable to nonstromatal species. In *Sordaria* there is no sign of an endogenous rhythm; the cycle is entirely dependent on the cycle of light changes in the day. Under conditions of illumination of 12 hours light (100 f.c.) and 12 hours dark in each day, spore discharge is at a low level during the period of darkness. With the onset of light, discharge rate rises to a maximum and thereafter falls, but not to the level of discharge in darkness. The change from light to dark always involves a rapid decline in rate. It has been shown that it is the short rays of light (400 to 500 *m*μ which are effective in stimulating discharge.

It should be remarked, however, that in the investigations using *Daldinia* and *Sordaria*, temperature has been kept constant and the water supply has been in no way limiting.

Mention has already been made of the periodicity in trapping of the spore of *Peronospora tabacina*. Waggoner and Taylor observed a maximum of spores in the morning, and associated this with the daily decrease of humidity in the early hours operating the hyposcopic mechanism of dispersal.

Work in Britain and in Nigeria shows that diurnal periodicity of spores in the air is the rule with such diverse types as the brand spores of *Ustilago*, the conidia of powdery and downy mildews, and the uredospores of rusts, although it remains to be seen if this is due to periodicity of spore liberation, to rhythm in the turbulent conditions of the air bringing spores to the spore trap orifice, or to a combination of both.

Much remains to be determined about the daily cycle of spore liberation, but from what is known the periodicities observed do not appear to have much selective value for the pathogen in the sense that the fungus achieves a greater infection as a result of restricting its spore output to certain times a day.

We are left with a general picture of spore liberation in which one of the principal features is the lack of exact knowledge. However, plant pathologists generally are becoming increasingly interested in the quantative aspect of dispersal, including the take-off, in the pathogens they study, and in the next decade we may hope to see the development of a much more detailed picture.

PART B
DISPERSAL BY AIR AND WATER—
THE FLIGHT AND LANDING

The problem of dissemination of plant diseases, whose propagules (such as fungal spores) cannot move by themselves, is more amenable to theoretical treatment than any other problem in phytopathology. This is especially true for air-borne pathogens. Once in the air, the way these propagules travel, the horizontal and vertical distance covered, and the time in transit all depend exclusively on external forces of a physical nature. Thus, the problem assumes a physical rather than a biological character.

A knowledge of these forces is a prerequisite for theoretical treatment of the agent's flight, some concepts in this area of phytopathology are unclear. Often, only two forces or effects of forces are considered, i.e., velocity of wind (horizontal aerial movement) and velocity of fall that the agent would have in calm air. Thus, for instance, Christensen (1942) calculated that a spore of *Ustilago zeae*, when brought 1 mile upward at a wind velocity of 20 m.p.h., can travel a distance of 2500 miles in about 9 days. Such calculations, however, have little practical and no theoretical value. Such an observation cannot explain how the spore got so high up. Since the wind acts horizontally and the velocity of fall vertically, no possibility of upward movement is considered; yet spores and pollen are found even in the highest atmospheric strata. First, one has to find out which external forces are responsible for transportation of the causative agents and how the individual force components affect the agent. Then a theory of agent flight can be established.

I. TRANSPORTATION FORCES

In discussing the determining factors for transportation of the agents through air, one need not consider molecular forces and electrostatic effects. Although they could exert an effect on the movement of the causative agents, they are so much overshadowed by other influences as to be negligible.

Gravitation affects each small particle in the air; together with buoyancy and internal friction of the air it acts constantly to give each particle its own falling velocity (in calm air). The horizontal movement of the air, called wind, is superimposed over this direction of movement. Finally, turbulence results in the vertical movement of air and contained particles.

The resultant of all these forces then determines what pathway the particles follow from take-off to landing. These forces and their effects will be examined individually and briefly.

A. Gravitation

According to Newton's law of gravitation, an attracting force k acts between two bodies with masses m_1 and m_2 that are at a distance r from one another:

$$k = G\frac{m_1 m_2}{r^2} \text{ [dyn]} \tag{1}$$

whereby G is the universal gravitation constant, which is independent of the state of both bodies. The terrestrial force of gravity is the resulting attraction between the mass of the earth and bodies found on or near it. This gravitation imparts to each body a downward accelerated movement during free fall. In the atmosphere the effect of gravitation is partly compensated by buoyancy and by the opposing force of internal friction of the air. The net effect of these forces, as mentioned, is the velocity of fall inherent in each particle.

B. Velocity of Fall

A sphere-shaped particle with radius r that moves in the (calm) air with velocity v experiences, according to Stokes' law, a resistance of the size

$$k = 6\pi\eta vr \text{ [dyn].} \tag{2}$$

The reference is valid in this form only under the condition that the radius of the particle is larger than the free path of the corresponding gas with viscosity η. Since the average free path in the air only measures 0.1μ and the diameter of spores generally is about 10μ, this condition can be considered as adequately fulfilled. Thus, as the particle is accelerated by gravitation (diminished by the insignificant amount of air buoyancy), there is an increase in frictional resistance. Thus, an equilibrium is established between the weight k' of the particle and the frictional resistance k so that the particle falls with constant velocity. Since the weight of the sphere-shaped particle is given by

$$k' = \frac{4}{3}\pi r^2 sg \tag{3}$$

where s represents the density of the particle and g terrestrial acceleration, fall velocity of sphere-shaped particles can be calculated from the equilibrium condition (buoyancy neglected).

$$\frac{4}{3}\pi r^2 sg = 6\pi\eta vr. \tag{4}$$

It is

$$v_k = \frac{2\ sg}{9\eta}\, r^2 \tag{5}$$

In phytopathology we are mostly concerned with agents that are not spherical, but ellipsoid, as is the case (at least approximately) in fungal spores. Falck (1927), who concerned himself very thoroughly with velocity of falling fungal spores in calm air, attempted to correlate experimentally the velocity of falling ellipsoid particles with the velocity of spherical particles of the same volume and found the relation

$$v_e = \frac{v_k}{2\sqrt{a/b}} \tag{6}$$

whereby v_e represents velocity of falling ellipsoid particles, v_s velocity of falling spherical particles of equal volume, and a and b the axes of the ellipsis. From equation (5)

$$v_e = \frac{2\ sg}{9\eta\ \sqrt[3]{a/b}}\, r^2 \tag{7}$$

is derived. When the volume of the spherical and ellipsoid particles are equal,

$$v_k = \frac{4}{3}\pi r^3 = \frac{4}{3}\pi ab^2 = v_e. \tag{8}$$

Radius r can be written as

$$r = \sqrt[3]{ab^2}. \tag{9}$$

Inserted into equation (7), the equation for velocity of falling ellipsoid particles is

$$v_e = \frac{2\ sg}{9\eta}\ \sqrt[3]{a} \times b\ \sqrt[3]{b^2}. \tag{10}$$

Thus, the velocity of falling ellipsoid fungal spores in calm air depends essentially only on the size relationship of axes a and b of the ellipsoid (s, g, and η can be considered constant in this observation) and establishes a physiological value determined by form.

The velocities of fall determined according to equation (10) by Schrödter (1954) for various groups of spores lie between less than 1 mm./sec. and a few cm./sec. This is in accord with reports by Ingold (1953), according to whom the very small spores of *Lycoperdon pyriforme* have the small fall velocity of 0.05 cm./sec., whereas the large conidia of *Helminthosporium sativum* fall 2.8 cm./sec.

The falling velocity (in calm air) of different propagules is an important characteristic in their dispersal. This fact is not always properly recognized. Although Gregory (1952) admits that the velocity of fall of spores has some influence on the total distance of dissemination, he neglects this influence as long as the spores can be found in the turbulent air movement. He thus overlooks the fact that the gravitational fall is always going on. Even in turbulent air when the net movement of the particle is upward, gravitational fall continues. Among other things, Fortak (1957) while discussing the problem of the extraterritorial sphere of influence of dust transportation from land to open sea, shows how greatly the fall velocity of the air-borne particle affects its dispersal. His results show a numerically strong dependence on the sedimentation parameter. Our discussion of the flight of the propagule with clarify even more the great significance of fall velocity in dissemination.

C. Horizontal Air Movement

Horizontal air movement is one of the most important factors in the dispersal of plant pathogens and of plants and seeds as well. The wind, a term mostly used for horizontal air movement, is considered most important in epidemiology.

Wind is moving air. The cause of this movement lies ultimately in temperature contrasts as they occur everywhere on the globe of the earth. Aside from gravitational fall, spores and other suspended particles move with the air. There is not just a horizontal air movement in the atmosphere. The paths of individual air particles as well as of the suspended matter are actually extremely entangled and "disorganized." The horizontal component of the air movement determines the direction and the velocity of movement of spores in air. The direction of spore dissemination and the velocity, which in turn affects the distance that the propagules travel, give the pathologist information as to where new disease outbreaks are likely to occur. This is why horizontal air movement assumes importance in general epidemiologic observations and examinations. From numerous works that have been concerned with this part of the problem only those by Newhall (1938), Oort (1940), Bonde and Schultz (1943), and Waggoner (1952) are mentioned as examples. The direction of the horizontal spread of agents is not only dependent on the general, i.e., average or moderately large wind direction, but also in great measure on the local regional relationships that affect the air current. A convincing examples for this is given by studies of Zogg (1949) about epidemiology, of *Puccinia sorghi*, conducted on the occasion of a maize rust epidemic in a river valley 100 km. long.

Under the influence of gravitation and wind alone (as horizontal air movement) a spore can never fly higher than its point of departure lies (with the exception of a forced upward movement of air while running over an obstacle). Within this framework very narrow limits are put on the dispersal of propagules. Observations prove, however, that spores rise high in the atmosphere and that dissemination over great distances is possible. Considerable height differences can be overcome by spores and a decisively significant vertical force seems to be present. This force lies in a property that characterizes all movements in the atmosphere, namely, in atmospheric turbulence.

D. Atmospheric Turbulence

If we study the flowing fluids with different density δ and diverse viscosity η in tubes of various diameters d, it can be seen that the stream is "laminar" in small velocities and "turbulent" in high ones. It can be further seen that the transition from laminar to turbulent state dose not occur in a gradual increase of velocity, but suddenly, i.e., in transgressing a limiting value of velocity. This limiting velocity v is not constant, however. Rather, the transition from laminar to turbulent flow occurs when a critical value of the so-called "Reynolds number" is exceeded. The Reynolds number R is given in

$$R = \frac{\delta}{\eta} \times vd. \tag{11}$$

In atmospheric currents the value of the length d is always so large that even when velocity v is low, the value of the Reynolds number is high. Thus, the laminar type of current practically does not occur in the atmosphere, certainly not in the layers that are significant for the disseminations of propagules. This fact is very important for all epidemiologic observations, since it dooms to failure every attempt to connect dissemination of pathogens with wind unless turbulence is considered.

The "dynamic turbulence" is closely connected with the so-called "shearing" between air strata of various horizontal velocities. These are always present in the atmosphere. We have only to imagine that the wind velocity near the earth's surface is smaller than at a higher altitude. Because of turbulence in air, air particles and suspended matter travel from one stratum into the other and an exchange takes place in a vertical direction (the horizontal turbulence exchange should not be considered here). This vertical transportation of air quanta as a result of dynamic turbulence disappears when the wind does not change with altitude because the above mentioned shearing then also disappears. However, this does not mean that the exchange in a vertical direction then disappears altogether. Only such turbulence is equal to zero as derives its energy from the current.

Another very important vertical change occurs, namely, the one that is correlated with the thermic increase of the atmosphere as well as with its thermic stratification and the heat supply of the earth. It is a known fact that warm air is lighter than cold air. An air quantum that is warmer than its surroundings has buoyancy (according to the principle of Archimedes) and moves vertically upward while another comparatively colder air quatum takes its place. Above all, the thermic differences of the ground are very significant for "thermic turbulence." Individual occurrences of a vertical exchange of greater or smaller turbulence bodies can barely be observed through dynamic turbulence. Only its total effect becomes visibly manifest and the occurrences of a thermic turbulence exchange can be observed in numerous individual manifestations. In addition to vibration of the air over an overheated surface the most marked manifestation is the appearance of large swelling clouds that can be especially seen during the height of summer, and into which great quantities of air ascent from the overheated earth surface. Naturally, the same amount of air descends again between the clouds. Significantly, the thermically caused upward movement of spore-laden air occurs quickly and over a relatively small land surface, while it subsides slowly and over great areas. Thus, Firbas and Rempe (1936) did not find the expected distribution according to size and fall velocity in pollens caught at a height of 2000 meters during an airplane flight. The strong and rapid anabatic wind stream seems to carry along the total mass of pollen bodies; there is no possibility that the differential size and fall velocity can have any effect on distribution. Night flights thus brought forth the originally expected results.

As far as epidemiology is concerned it is unimportant whether these occurrences in the atmosphere are "thermic" or "dynamic" turbulence. Their effect is the same, i.e. there is a possibility of a vertical agent transfer through vertical exchange of air quanta. Through the exchange calculations of Schmidt (1925) all these turbulence reactions were organized into one uniform scheme. The surpassing significance of the vertical mass exchange for epidemiology of plant diseases, shown by Schrödter

(1954), requires that a short abstract of the theory of vertical mass exchange be given before the theory of the flight of the pathogen. In this way the problem of agent dispersal can be fully understood.

During the exchange process all those properties of air masses are exchange that had remained preserved during vertical movement i.e. the ascending or descending air quanta retain their respective property until they are mixed with the surrounding air—and in this mixture they give up their respective surplus property to the surrounding air or they cover the deficit with the surplus from the surrounding air. Thus, the organic or inorganic particle content of the air belongs to the interchangeable properties. The number of particles of an air quantum remains practically unchanged during sufficiently rapid vertical movement as long as there occurs no mixture with other air masses.

Imagine a horizontal surface at some place in the atmosphere that, because of turbulence, is constantly carried upward or downward by air quanta. Thus it is unimportant whether the surface is quiet or whether it moves with the average current of the atmosphere. After a sufficiently long period of time, because of the continuity condition, the sum of the air quanta ascending through a given area per unit time should equal the sum of the air quanta descending through this area per unit time. The total "mass flow" M of the exchange is thus given by

$$\frac{1}{2} M = \Sigma\, m_u = \Sigma\, m_o \tag{12}$$

when we designate the quanta ascending through the imaginary plane surface with the subscript u and the descending quanta with the subscript o. If we then consider the number of spores contained in an air quantum as an interchangeable property, each ascending quantum m_u contains the spore quantity $m_u s_u$ and each descending quantum m_o the quantity $m_o s_o$. The total transfer that occurs through the surface unit in the time unit is thus given by

$$S_o = \frac{1}{ft} (\Sigma\, m_u S_u - \Sigma\, m_o s_{o)} \tag{13}$$

whereby f represents the plane surface and t the time.

It seems at first glance that $S = 0$ because of equation (12). This, however, is not so. According to equation (12) M only represents the flow of mass, and nothing is said about the property that is transferred with the mass. As we know, the spore content of air depends on height. Near the surface, where spores develop and are ejected, the number is naturally larger than in the higher strata of the atmosphere. This fact is expressed in a certain vertical distribution of the spore content of air. If we imagine that the spore content contains s-1 units above the surface and $s + 1$ units below the surface (in other words it changes by 2 units with height) and if we shift one quantum from below to above and as a replacement one quantum from above to below, we gain above $(s + 1) - (s - 1) = + 2$ units and lose below $(s - 1) - (s + 1) = - 2$ units. Although the condition of equation (12) is fulfilled, actually a flow of property s took place, i.e., s is not equal to zero. S equals zero only when it does not depend on altitude. The change of property s with altitude, i.e. its vertical distribution, is the decisive factor.

If we count altitude z, within which no mixture occurs, from the imaginary surface f on, whereby above f, $z > 0$, we can describe the distribution of s near the surface f through development in a series. Using only the first member of this series, we have

$$s = s_f + \frac{\partial s}{\partial z} \times z + \frac{\partial^2 s}{\partial z^2} \times \frac{z^2}{2} \tag{14}$$

whereby s is the sum of s_f, the value of s in the surface f plus the first and second derivatives of s with respect to z taken in surface f. For a first approximation we may ignore the second derivative and we may write for values s_u and s_o in equation (13)

$$s_u = s_f + \frac{\partial s}{\partial z} \times z_u \tag{15}$$

$$s_0 = s_f + \frac{\partial s}{\partial z} \times z_o.$$

If, with equation (12) in view, we substitute the values of equation (15) into equation (13) we have

$$S = \frac{1}{ft} (\Sigma\, m_u z_u - \Sigma\, m_o z_o) \frac{\partial s}{\partial z} \tag{16}$$

for the spore flow through the imaginary surface. The parenthetic expression is always negative since we consider z_u as negative and z_o as positive. According to calculation, however, this parenthetic expression represents the sum of all air quanta m that pass through the imaginary surface, either upward or downward, each multiplied by the distance z travelled without the mixture ("mixing length" in the sense of Prandtl) at which it was from the surface. It is thus

$$S = -\frac{\Sigma\, mz}{ft} \times \frac{\partial s}{\partial z} \tag{17}$$

or

$$S = -A \times \frac{\partial s}{\partial z} \tag{18}$$

whereby we obtain A, the exchange coefficient:

$$A = \frac{\Sigma\, mz}{ft}\left[\frac{gm}{cm.\, sec.}\right] \tag{19}$$

With equation (18) the flow of the interchangeable material, in this instance the spore content of air, is represented as the product of two factors, of which one ($\partial s/\partial z$) depends only on the vertical distribution of the interchangeable property, the other (A) only on the movement processes, i.e., on the mass exchange. Hence the designation "exchange coefficient," which measures the turbulent movement processes in the atmosphere. Its special value for our observations lies primarily in the fact that according to equation (19) the exchange coefficient A is independent of the exchanged property. Numerically, the exchange coefficient in the atmosphere is subject to considerable changes. Its smallest value is near the soil, because naturally the soil, as a firm boundary surface, impedes any vertical movement. The exchange coefficient thus depends on altitude. In orders of magnitude A has the value of 1 gm. cm.$^{-1}$ sec.$^{-1}$ from 1 to 10 meters altitude, the value of 10 at 10 to 100 meters altitude, and of 50 at 100 to 500 meters altitude. In single instances A can take on values of 200 or more.

In epidemiology the exchange coefficient characterizes the total turbulent processes in the atmosphere which are responsible for the transfer and dispersal of propagules. To determine whether an effect of dynamic or thermic turbulence is involved in vertical transfer is no longer necessary. What deciding role turbulent mass exchange plays in the spread of the pathogen must yet be demonstrated.

II. THE FLIGHT

The forces having an effect on the particles in the air have been discussed. The theory of agent flight should show how the transfer in the air as well as the dispersal of propagules occurs. The term "dispersal" is defined here to mean the overcoming of distance in space from the time of the last departure to the first landing. The problem is considered only at the x–z plane of a rectangular coordinate system, in which the x-axis lies in the average wind direction and the z-axis vertically, i.e., only two-dimensionally. Such a precise definition of the term "dispersal" is to be understood geometrically and not biologically. The concept of dispersal as used elsewhere in the literature has several meanings, depending on whether the flight plane, the spore concentration in a unit volume of air, or the infection possibility is considered. This ambiguity not only leads to lack of clarity, but can also result in false notions and conclusions. Thus, in his observations Gregory (1952) follows not the path of the spore cloud, but the changes of concentration per air volume along this path as a consequence of turbulence. These, however, are two completely different problems. Contrary to all expectations he concludes that the spores cannot be carried far from the source of infection but that under normal turbulent conditions 99.9% of the spores are deposited within the first 100 meters from the source. He concludes this from theoretical assumptions based on the equation by Sutton (1947) about turbulent diffusion and on experimental observations, from the viewpoint of the change in concentration of spores and its effect on the possibility of infection. This has nothing to do with the distance that the mass of spores can travel.

Epidemiologically, both problems are equally important: the absolute distance as well as the maximum distance in which a significant concentration of spores is still present. The concept "dispersal," however, is used in this part of the study exclusively for the first of both these questions and is thus completely unequivocal.

Schmidt (1925) was already concerned with the theoretical aspect of the spread of plant seeds through turbulent air currents. More recently Rombakis (1947) tackled this problems again and proved that the results obtained by Schmidt (1925) could be improved in various ways. The elegant solution of the problem developed by Rombakis (1947) was used by Schrödter (1954) for the problem of spread of disease-producing agents. The theory of spore flight, described below, is based on the studies of Rombakis (1947). The somewhat extensive derivation of the formulas leads to an understanding of the problems of line of flight, range of flight, and duration of flight, which can then be applied to problems of epidemiology.

A. Line of Flight

The studies by Schmidt (1925) define as "average dispersal" of particles that limit in which 99% of all disseminated seeds have again reached the earth surface. Had the percentage been set at 80%, based on purely biological considerations, the dispersal would be but 13% of the values calculated by Schmidt (1925). Although he did not consider the effect of gravitation in the basic differential equation, he added it later to the solution. The theory given by Rombakis (1947), on the other hand, is based on considerations to be applied here to fungal spores.

The local change of spore thickness s with the time t has to be equal to the convergence of spore flow w, i.e., it has to be

$$\frac{\partial s}{\partial t} = \frac{\partial w}{\partial z} \tag{20}$$

The spore flow w is made up of the current

$$w_1 = \frac{A}{\delta} \times \frac{\partial s}{\partial z} \tag{21}$$

that is caused by exchange, corresponding to equation (18) in which δ is air density, and of the current

$$w_2 = -cs \tag{22}$$

caused by fall movement under the influence of gravitation, whereby c means the fall velocity of spores in calm air. The total spore current is thus given by

$$w = w_1 + w_2 = -\frac{A}{\delta} \times \frac{\partial s}{\partial z} - cs. \tag{23}$$

Substituting this expression into equation (20) gives the complete differential equation for dispersal of spores

$$\frac{\partial s}{\partial t} = \frac{A}{\delta} \times \frac{\partial^2 s}{\partial z^2} + c\,\frac{\partial s}{\partial z} \tag{24}$$

which, contrary to Schmidt (1925), includes the fall effect due to gravitation. On condition that at time $t = 0$ and at place $x = 0$, $z = 0$ a number N of spores is dispersed in the open space $z > 0$, the number n' of spores, found above z at time t, is given by

$$n' = \int_z^{\infty} n\,dz = \frac{2N}{\sqrt{4\pi a t}} \times \exp\left(-\frac{c^2 t}{4a}\right) \int_z^{\infty} \exp\left(-\frac{z^2}{4at} - \frac{c}{2a} \times z\right) dz \tag{25}$$

where $a = A/\delta$.

The following definition of "probable flight line" leads to an unequivocal solution that is independent of any arbitrary limitation: A point P at the altitude z at time t should be considered a point of probable flight line when it is also probable that a spore is found above as well as below this point. This means that the line is determined, above as well as below which 50% of all spores are found and dispersed. This percentage, contrary to that chosen by Schmidt, is not an arbitrary number but the condition for the equation of two probabilities. The probable line of flight is thus defined by

$$\frac{n'}{N} = \frac{1}{2} \tag{26}$$

and from equation (25) it follows that

$$\frac{1}{2} = \frac{2}{\sqrt{4\pi at}} \times \exp\left(-\frac{c^2t}{4a}\right)\int_z^{\infty} \exp\left(-\frac{z^2}{4at} - \frac{c}{2a} \times z\right) dz. \tag{27}$$

The value of an integral corresponding to that in equation (27) can be written in general form as

$$\int_{\alpha}^{\infty} \exp\,(-\xi^2 - 2\beta\xi)\; d\xi = \frac{1}{2}\sqrt{\pi}\,(\exp \beta^2)\,[1 - \Phi\,(\alpha + \beta)] \tag{28}$$

In equation (28) the expressions α and β correspond to

$$\alpha = z/\sqrt{4at}$$

and to

$$\beta = \sqrt{c^2t/4a}$$

in equation (27). Thus equation (27) can be rewritten to give

$$\frac{1}{2} = 1 - \Phi\left(\frac{z}{\sqrt{4at}} + \sqrt{\frac{c^2t}{4a}}\right) \tag{29}$$

or

$$\Phi\left(\frac{z}{\sqrt{4at}} + \sqrt{\frac{c^2t}{4a}}\right) = \frac{1}{2} \tag{30}$$

This transcendental equation has the root

$$\frac{z}{\sqrt{4at}} + \sqrt{\frac{c^2t}{4a}} = 0.4769 \tag{31}$$

from which it follows that

$$z = 0.4769\,\sqrt{4at} - ct. \tag{32}$$

In a coordinate system, a wind of average velocity U travels the distance x in time t, or

$$x = Ut \tag{33}$$

and we can write equation (32) as

$$z = 0.4769\,\sqrt{\frac{4Ax}{\delta U}} - \frac{c}{U} \times x. \tag{34}$$

This is the equation of the probable flight line.

The question about the shape of the flight line of spores can be answered when equation (34) is squared and transposed to the form

$$\frac{c^2}{U^2} \times x^2 + 2\frac{c}{U}xz + z^2 - (0.4769)^2 \times \frac{4A}{\delta U}x = 0. \tag{35}$$

An equation of the second degree of general form

$$a_{11}x^2 + a_{22}z^2 + 2a_{12}xz + 2a_{11}x + 2a_{23}z + a_{33} = 0 \tag{36}$$

describes a parabola of which the determinant is

$$D = \begin{vmatrix} a_{11}a_{12} \\ a_{12}a_{22} \end{vmatrix} = 0 \tag{37}$$

Since, in the case of equation (35) and (36), $a_{11} = (c/U)^2$, $a_{12} = c/U$ and $a_{22} = 1$, we may write the determinant

$$D = \begin{vmatrix} \left(\frac{c}{U}\right)^2 & \frac{c}{U} \\ \frac{c}{U} & 1 \end{vmatrix} = 0 \tag{38}$$

which fulfills the condition given in equation (37). The line of flight is thus a parabola. Since the inclination of the axis of a parabola is given by

$$\tan(2\alpha) = \frac{2\tan\alpha}{1 - \tan^2\alpha} = \frac{2a_{12}}{a_{11} - a_{22}} \tag{39}$$

the inclination of axis for the line of flight parabola is

$$\tan\alpha = -\frac{c}{U} \tag{40}$$

The line of flight of spores is a parabola. The inclination of its axis is the vector determined by the velocity of fall of spores in quite air and the average wind velocity (compare with Fig 4.17).

As mentioned previously, the course of a single particle is unusually complicated and practically not reproducible because of the "disorganized" movement due to turbulence. The probable flight line does not describe the course of an individual practice, which is per se epidemiologically uninteresting. Rather, the probable flight line is a curve representing the average path of ejected spores, 50% of which occur below and 50% above this line as the mass of spores expands under the influence of wind, turbulent mass exchange and, fall caused by gravitation. We can best characterize this probable flight line by describing it as the course of the centre of gravity of the transferred spore cloud. Figure 4.17 shows how great the significance of fall velocity is in this problem. Lines of flight are compared, using varying values of c with corresponding

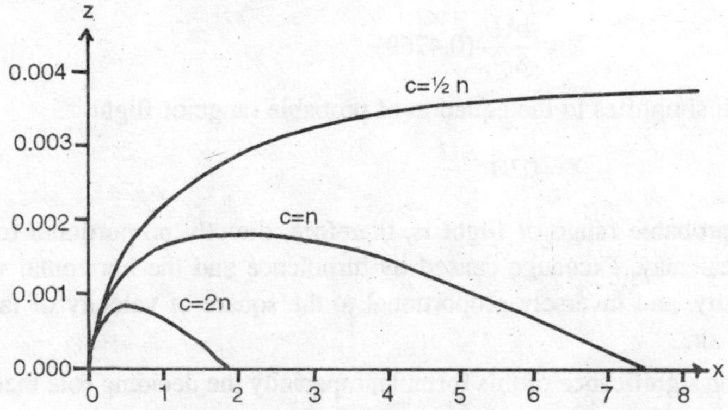

Fig. 4.17. Form of flight line with varying velocity of fall (c), equal mass exchange and equal wind velocity (z-axis increased 1000 times).

values of A and U. Velocity of fall must be considered in relation to the turbulent mass exchange when discussing the problem of dispersal. Thus, if fall is neglected, no solution of the problem can be reached.

B. Range of Flight

Especially interesting from an epidemiological point of view is the question of how far the spores of pathogens can fly. This can be answered from theory when an unequivocal definition is also given for the range of flight while developing the theory given above, Rombakis (1947) deduces this definition from the one of probable flight line. The "probable flight range X" is, accordingly from the distance of the source of spores to where the probable line of flight again cuts the earth surface. Between the spore source $x = 0$, $z = 0$ and this cutting point ($x = X$, $z = 0$) only 50% and not 99% of the spores (as per Schmidt's definition) get to the ground. According to the definition of flight line it is also probable that a spore flies to the point $x = X$ or travels an even greater distance. If, however, we consider a point $x = X + m$, we know that more than half of the spores are disseminated up to this point. When we consider a point $x = X - m$, we know that more than half of the spores disperse above this point.

When, according to this definition in equation (34), we say $z = 0$ and $x = X$, the determining equation for the probable range of flight is

$$0 = 0.4769\sqrt{\frac{4A}{\delta U}}\,X - \frac{c}{U}X \qquad (41)$$

or, in terms of X,

$$X = \frac{4AU}{\delta c^2}(0.4769)^2 \tag{42}$$

which simplifies to the equation of probable range of flight

$$X = 0.91\frac{AU}{\delta c^2} \tag{43}$$

The probable range of flight is, therefore, directly proportional to the vertical mass exchange caused by turbulence and the horizontal wind velocity, and inversely proportional to the square of velocity of fall in calm air.

The significance of this formula, especially the deciding role that fall velocity plays is clearest in some examples given in Table 4.1, when the ranges of flight are compared one with another in these four examples.

TABLE 4.1

Flight Range (X,) With Different Values of Mass Exchange (A), Wind Velocity (U), and Velocity of Fall (c)

	A	*U*	*c*	*X*
	gm./cm. sec.	*m./sec.*	*cm./sec*	*km.*
1.	10	4	2	7.6
2.	20	4	2	15.2
3.	20	8	2	30.3
4.	20	8	1	121.3

As seen in examples 1 and 2 of that table, the range of flight is doubled when the mass exchange is doubled. The same holds true for doubling the horizontal wind velocity (examples 2 and 3 of the table). If, on the other hand, the velocity of fall is cut in half (examples 3 and 4), the range of flight becomes fourfold, since the fall velocity is squared in the formula. Thus, contrary to Gregory (1952) and Ingold (1953) the velocity of fall is an extremely important factor in determining the range of flight and cannot be neglected in the problem of dissemination. This can be also seen clearly in Fig. 4.17. If the curve given in Fig. 4.17 is plotted on coordinates where the ratio of the scales of ordinate; abscissa ratio is 1 : 1 (rather than 1 : 1000), we obtain a flat, slowly ascending curve that appears almost straight even at a small distance from the origin (Fig. 4.18). Even in small turbulence the effect of fall is almost imperceptible. This may be the reason for underestimating its significance. However, Rempe (1937) showed the effect of gravitation on the distribution of pollen sizes at various altitudes.

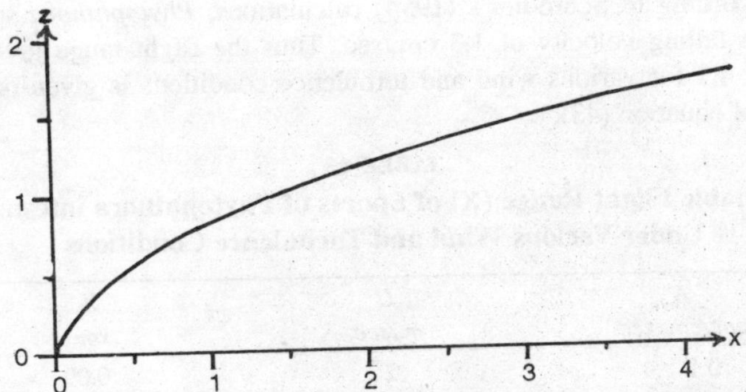

Fig. 4.18. Form of flight line in a mass exchange of 10 gm./cm. sec., a horizontal wind velocity of 1 m./sec. and a fall velocity of 1 cm./sec. (with an equalized standard distribution of co-ordinate axes).

The range of flight can be great even under normal wind and turbulence, as is shown in Table 4.1. How great these can be was calculated by Schrödter (1954) for various spore sizes (Table 4.2). Although large spores can fly as far as a few kilometers, they still fall on the ground within visible distances. The very small spores, on the other hand, should be described as "suspension particles" within the air plankton.

TABLE 4.2

Probable Flight Range of Spores of Different Sizes with a Wind Velocity of 2 m./sec. and a Mass Exchange of 10 gm./cm. sec.

Spore size (length/width) μ	*Velocity of fall (cm./sec.)*	*Flight range (km.)*
Small spores (5 : 3)	0.035	12,400
Medium spores (14 : 6)	0.138	800
Large spores (20 : 16)	0.975	16

In numerous epidemiological studies attempts have been made repeatedly to solve the problem of how far spores of pathogens can fly. The question often assumes great importance, especially in studies on the epidemiology of *Phytophthora infestans*. Without question, spore flight has an important significance in the difficult problem of forecasting epidemics. Schrödter (1954) has applied the theory on the dissemination of *Phytophthora infestans* to the epidemiology of potato blight and has compared predictions based on the previously given theory with practical observations.

According to Schrödter's (1954) calculations, *Phytophthora* spores have a falling velocity of 1.3 cm./sec. Thus the flight range given in Table 4.3 for various wind and turbulence conditions is given on the basis of equation (43).

TABLE 4.3

Probable Flight Range (X) of Spores of Phytophthora infestans Under Various Wind and Turbulence Conditions

A gm./cm. sec.	*U* m./sec.	X km.
0.1	2	0.09
1	4	1.8
10	6	27
20	8	72
50	10	225

It is difficult to carry out direct observations on how far spores of *Phytophthora infestans* fly. Generally, therefore, the distance was judged from observed infections in the respective studies. Such data should be considered as underestimates of the distance spores can fly, because such observations tell only how far they can fly in concentrations sufficient to produce an infection while retaining their germinative capacity. Such observations also assumed that environmental conditions favor the development of the organism at the spot where the spores reach the ground. While these matters are of epidemiological interest, they have nothing to do with the actual flight range. But comparison of the observations with theory is possible only under these conditions.

That spores of *Phytophthora infestans* can be carried very far by the wind because of their light weight was stressed by Fischer and Gäumann (1929). Hänni (1949) also concludes that they can be carried over vast distances, but assumes this to occur only after the passage of considerable time, and direct dissemination by the wind to occur only over relatively short distances. While Bonde and Schultz (1943) established flights of 200 meters, Hänni (1949) observed flights of 1 km. In spores caught at various points over a large area Raeuber (1957) could establish that the rhythm of the daily number caught remains the same over several kilometers, from which fact a slight mobility and a corresponding broad dispersal can be concluded. Thomas (according to van der Zaag, (1956) could catch spores at 4.8 km., while Hyre (cited by van der Zaag, 1956) caught them even at a 14 km. distance from the spore source. Van der Zaag (1956) concludes from observations on an island several kilometers from land that spores can fly at least as far as 11 km. without losing their capacity to germinate. From his observations Godfrey reports distances of 24 km., while Harrison suggests even those of 48–64 Km. (Table 4.4).

TABLE 4.4

Observed Flight Ranges from the Literature of Spores of Phytophthora Infestans

Observations of author (Reference)	*Flight range (km.)*
Bonde and Schultz	0.2
Hänni	1
Raeuber	> 2
Thomas	48
van der Zaag	> 11
Hyre	14
Godfrey	24
Harrison	48 – 64
Fischer and Gäumann	very far

If values given in Table 4.4 are compared with data in Table 4.3 we note, notwithstanding the restriction mentioned above, that the observed flight ranges are consistent with those calculated on the basis of theory. Although data in Table 4.4 vary from 200 meters to more than 60 km., it would be useless to dispute over what is the true flight range. Such differences can be caused by differences in the degree of turbulence of the atmosphere and by differences in horizontal wind velocity. At any rate, comparison of Table 4.3 and 4.4 shows that theory and observation agree.

C. Height of Flight

During their flight through the air, propagules have frequently been found at high altitudes, even in the highest strata of the atmosphere. Thus we may ask what is the ceiling of the flight line parabola. The "maximum probable flight altitude" can be derived from the data given so far: by analogy with the previous definitions it is the maximum height that the center of gravity of the spore cloud can achieve. Thus, although one spore can fly above this altitude, less than 50% of spores are dispersed above this point of the maximum probable flight line.

According to this definition (Rombakis, 1947) the condition $dz/dt = 0$ is valid for the highest point of probable flight line. If we use this condition in equation (32), we get

$$\frac{dz}{dt} = 0.4769\sqrt{4a} \times \frac{1}{2\sqrt{t}} - c = 0. \qquad (44)$$

Solving this equation for *t*, we get

$$t = \frac{(0.4769)^2}{4} \times \frac{4a}{c^2} \qquad (45)$$

If we now substitute the value of *t*, [equation (45)] in equation (32) under the condition $dz/dt = 0$, we obtain the highest point of the flight line:

$$z_{max} = 0.4769 \times 4a \times \frac{0.4769}{2} \times \frac{4a}{c} - c \times \frac{(0.4769)^2}{4} \times \frac{4a}{c^2}$$

$$= (0.4769)^2 \times \frac{a}{c} \qquad (46)$$

and with it, because $a = A/\delta$, the equation of maximum probable flight altitude

$$z_{max} = 0.2274 \times \frac{A}{\delta c} \qquad (47)$$

The maximum probable flight altitude is accordingly directly proportional to the size of turbulent mass exchange and inversely proportional to velocity of fall. It ia independent of horizontal wind velocity.

Here for the first time is a definition of the size of dispersal, completel'' independent of wind as horizontal air movement. Naturally, the independence is not so remarkable per se because the horizontal wind velocity, as we have seen with analysis of the effect transportation forces, contributes no component to a vertical movement. Thus we derive from the equation that there is an epidemiologically specially significant dimension in the question of wind dispersal which is independent of the wind as such but entirely dependent on turbulence.

Since the velocity of fall is not squared in equation (47) as it is in equation (43), is effect on the altitude of flight, considered numerically, is smaller than on the range of flight. However, considering as a ratio the extent of the horizontally in relation to its vertical thickness we can visualize the atmosphere. From this point of view, the effect of velocity of fall must be at least as great on flight altitude as on flight range.

The flight altitudes that can be attained by spores in various mass exchanges and at different velocities of fall are given in table 4.5, and likewise, the maximum probable flight altitude for spores of *Phytophthora infestans* calculated from equation (47). Apparently spores, especially small ones, can attain considerable altitudes.

The accuracy of the theory can be tested by the results of catching spores during an airplane ascent. As Table 4.5 shows, we would expect spores to be distributed in altitude according to their size or velocity of fall. The pollen studies already mentioned by Rempe (1937) partly show the actually expected result, and indeed, show it with a single kind of pollen. Thus, for instance, during one night *Betula* pollen, caught at around 1000 meters altitude had an average diameter of 23μ, and that caught near the ground had an average diameter of 27 μ. During the day, however, pollen of all kinds and sizes was uniformly distributed at all altitudes. At a height of 2000 meters Firbas and Rempe (1936) did not find the expected distribution of pollens according to size and fall velocity. As far as *Phytophthora* is concerned, Hänni (1949) describes observations suggesting that spores have risen to altitudes of several hundred meters. According to Table 4.5, however, the maximum probable flight height under the turbulence occurring near the soil is likely to be less than 100 meters. Apparently there are discrepancies between theory and observation. Hänni's (1949) studies, however, clarify the apparent difficulty. These studies were carried out in mountain areas where the turbulence is characteristic but differs from that encountered in planes. Orographically conditioned vertical currents could develop, amounting to as much as 200 gm./cm. sec. or more in a small area and for a limited period of time. Under these conditions an ascent of spores of *Phytophthora infestans* to several hundred meters in altitude is not a problem. An explanation for the observations by Firbas and Rempe (1936) has been given in discussing turbulence as a transportation force. The strong vertical air movements, as they are known from the huge swollen clouds, allow the value *A* in formula (47) to increase markedly. Then the effect of velocity of fall on the size distribution of particles disappears for a limited time in a limited volume. Stakman et. al. (1923), in spore-trapping experiments in airplanes, found no spores. This indicates that these spores normally reach heights of less than 100 meters, as Table 4.5 shows. The results by Raeuber (1957), who caught spores at trapping stations at several locations near the ground as well as at an altitude of 22 meters, should also be used as confirmation of theoretical results. Not only is the rise and fall of the numbers caught daily at an altitude of 22 meters analogous to that occurring at ground stations, but also the total number of spores caught at this altitude was one-third of those caught at the normal ground stations directly below and one-tenth of the spore number caught at the true source of spores.

TABLE 4.5
Variation in Maximum Probable Flight Altitude of Spores of Differing Size with Vertical Mass Exchange

Spore size (length/width) (μ)	Fall velocity (cm./sec.)	*Maximum flight altitude in m. with a mass exchange A*			
		10	*20*	*50*	*100*
			(gm./cm. sec.)		
Small spores (5 : 3)	0.035	541	1082	2705	5410
Medium spores (14 : 6)	0.138	137	274	685	1370
Large spores (22 : 16)	0.975	19	38	95	190
Spores of *Phytophthora infestans*	1.3	15	29	73	145

Thus we can determine the probable flight altitude of the spores and see that theory and observation substantially agree with one another.

D. Duration of Flight

The question of how long the time is from the take-off of spores to their landing again has seldom been asked. The flight time is difficult to observe and can scarcely be determined experimentally. To be sure, calculations have been made, such as those by Christensen (1942), who estimated that a spore of *Ustilago zeae*, flying 1 mile above the ground, would come back to the ground only in 9 days. Such a calculation considers only the downward motion under the influence of gravitation and not the vertical movement resulting from the influence of turbulence. The question of duration of flight, then, can only be answered theoretically and one should refer to the derivation by Rombakis (1947). For the sake of completeness reference should be made to Schmidt's (1925) treatment of this problem.

Here, too, the duration of flight is understood as a "probable flight duration" since it naturally follows from definitions for flight line, flight range, and flight altitude. If we designate the duration of flight with z, we can proceed from the simple condition that $t = \tau$ must be at the place of $z = 0$, $x = X$, where the spore lands. Then, however, according to equation (32)

$$0.4769\sqrt{4a\tau} - c\tau = 0 \tag{48}$$

or after squaring,

$$c^2\tau^2 = (0.4769)^2 \times 4a\tau \tag{49}$$

when

$$\tau = (0.4769)^2 \times \frac{4a}{c^2} \tag{50}$$

and substituting A/δ for a, obtain, as equation of probable duration of flight,

$$\tau = 0.91 \times \frac{A}{\delta c^2} \tag{51}$$

We obtain the same result by dividing equation (43), i.e., the equation of probable flight range, by the horizontal wind velocity U.

The probable duration of flight is thus directly proportional to the vertical mass exchange and inversely proportional to the square of the velocity of fall. It is independent of the horizontal wind velocity.

Again we have a significant quantity for dispersal that does not depend on the wind as such. The duration of stay in the air is determined only by the vertical movement components, to which horizontal wind velocity does not contribute anything. From equations (47) and (51) we see that the maximum probable flight altitude is achieved in time $t = \frac{1}{4}\tau$, and the curve of the flight line has a steep ascent and a flat descent.

Table 4.6 gives the values of the flight duration under various turbulence conditions for various spores sizes, to which the respective values for spores of *Phytophthora infestans* are added.

TABLE 4.6

Probable Flight Duration of Spores of Various Sizes Under Varying Vertical Mass Exchange

Spore size (length/width)	*Fall velocity cm./sec.*	*Flight Duration under mass exchange A*			
		10	*20*	*50*	*100*
Small spores (5 : 3)	0.035	72 days	144 days	1 year	2 years
Medium spores (14 : 6)	0.138	5 days	9 days	23 days	46 hours
Large spores (22 : 16)	0.975	2¼ hours	4½ hours	11 hours	22 hours
Spores of *Phytophthora infestans*	1.3	1¼ hours	2½ hours	6¼ hours	12½ hours

We can see, therefore, that the flight in the air can last for 1 hour or 1 year, depending on size of spores and on turbulence. From Schrödter's (1954) calculations we see that the flight duration at small exchange values near the ground can be but a few minutes for large spores. On the other hand, a tiny spore 4μ in length and 1μ in width has a velocity of fall of 0.006 cm./sec. and can remain in the air more than 33 years at an average mass exchange of 50 gm./cm. sec. From a practical point of view such spores should only be considered as suspension particles for which the probability of reaching ground is very small, unless another external circumstance provides a back transfer to the earth's surface. The duration of stay in air as computed from equations cannot be checked by observation.

The duration of flight is significant from an epidemiological point of view in connection with the problem of viability of spores. Consider the spores of *Phytophthora infestans* as an example. According to a short summary by Raeuber (1957) the spores retain their vilability in dry air for a very limited time. When, on the other hand, we conclude from data in Tables 4.3 and 4.6 that these spores can cover a distance of 72 km. in only 2½ hours at a mass exchange of 20 gm./cm. sec. and a wind velocity of 8 meters/sec., we understand it to be a broad dispersal not only of spores, but also of infection. Because such conditions are fulfilled mostly in windy and rainy weather (i.e., under high atmospheric humidity) one can hardly count on a loss of viability in so short a time. In connection with the viability of propagules the duration of flights is thus also an epidemiologically important problem, about which theory can give adequate information.

Rombakis (1947) himself points to the fact that objections could be raised against the "exchange theory" when the particles to be transported leave their source individually, one after the other, and not as a group in large concentration at one time. Even under these two conditions identical results can be obtained by deductions from statistical physics.

III. CONCENTRATION IN THE AIR

The second important problem to be tackled deals with the variation in number of propagules per unit volume of air. We known from experience that the number of spores contained in a cubic meter of air changes with altitude. We know from spore-trapping experiments that the spore concentration is largest on the ground and that it decreases with altitude. From the theory of vertical mass exchange we know further that it is this vertical change of the spore content which should be considered a property of the air that elicits the "flow of property" in a vertical direction. Such a mass exchange occurs not only vertically, but also horizontally.

Thus we are dealing with occurrences that are very similar to the diffusion of gases. Diffusion is known to be a direct consequence of molecular movement, i.e., a compensation of density differences due to the random character of molecular movement. In a turbulent mass exchange we also have analogous disorganized movement and can consider the occurrences as a kind of diffusion in which, instead of molecules, larger air quanta are involved. The effect of such a turbulent diffusion is readily observed, e.g., the spread of a trail of smoke from a factory chimney. Like the trail of smoke from the factory chimney, the spore cloud coming from an infection source is dispersed in a horizontal and vertical direction, and as the cloud increases in volume, the spore concentration must become less and less.

A theoretical treatment of diffusion of small particles, emitted from a point source into a turbulent medium, was carried out by Ogura and Miyakoda (1954). Edinger (1955) also concerned himself with the dispersal by turbulent diffusion of particles too large to participate in Brownian movement. Turbulent diffusion in the air stratum near the ground was treated in detail by Sutton (1953). In the spread of plant pathogens the air stratum near the ground is the true place of observation. Therefore, the following is based on Sutton's presentation.

A. Turbulent Diffusion

The theory of turbulent diffusion in relation to the present problem contains a series of mathematical difficulties. Contrary to the problem of spore flight, which is concerned only with vertical mass exchange, the present problem is concerned with vertical and with horizontal mass exchange as well. A complete derivation will be sacrificed in this chapter, and instead a description of what is necessary for understanding of the basic equation for concentration change will be given. One should refer to Sutton (1953) and the literature cited for further details.

We limit ourselves to diffusion from a point source and use a coordinate system with the x-axis in the direction of the average wind, the y-axis at a right angle to the average wind, and the z-axis vertical. Thus at $z = 0$ we have the earth's surface. The average wind is considered to be steady and dependent only on altitude, so that the conditions $u = u(z)$, $v = w = 0$ are valid for the average velocity components in the direction of the axes. In addition, we assume that only small temperature gradients are present, that the ground neither absorbs nor emits the propagules being dispersed, and that during diffusion no propagules are deposited from the cloud. The particles are assumed to be so small as to have negligible falling motion. Although velocity of fall is significant for dispersal, we can, for the time being, assume that only short distances are under consideration and that the effect of fall is a needless complication to understanding the basic principle. A theoretical exposé that takes. sedimentation into consideration, as in agent flight, is given later in another context (see Section V).

The problem is the solution of the differential equation for diffusion, which is given in a general form by

$$\frac{ds}{dt}=\frac{\partial}{\partial x}\left(a_x\frac{\partial s}{\partial x}\right)+\frac{\partial}{\partial y}\left(a_y\frac{\partial s}{\partial_y}\right)+\frac{\partial}{\partial z}\left(a_z\frac{\partial s}{\partial z}\right) \tag{52}$$

in which s is the concentration in the x,y,z directions at time t, and a_x, a_y and a_z represent the parameters of diffusion in the direction of the coordinate axes. This equation is related to equation (24) for the case $c = 0$, i.e., the equation (24) is a special case of equation (52), when $a = A/\delta = \text{const.}$, and is the vertical component of this diffusion. If we consider a continuous point source, then neglecting the term

$$\partial/\partial x\,(a_x\,.\,\partial s/\partial x)$$

we have

$$u_{(z)}\frac{\partial s}{\partial x}=\frac{\partial}{\partial y}\left(a_y\frac{\partial s}{\partial y}\right)+\frac{\partial}{\partial z}\left(a_z\frac{\partial s}{\partial z}\right) \tag{53}$$

The difficulties connected with the solution of this differential equation were overcome by Sutton (1953) with the help of Taylor's theorem.

If σ is the standard deviation of distances, travelled at a time T by particles originally concentrated in the x-z plane, then we have

$$\sigma^2 = 2w'^2\int_0^T\int_0^t R\,(\xi)\,d\xi\,dt \tag{54}$$

in which $R\,(\xi)$ is the correlation coefficient between eddy velocity (w') at time t and $t+\xi$. According to Sutton (1953) $R\,(\xi)$ is given by expression of the form

$$R_z\,(\xi)=\left(\frac{\eta}{\eta+u'^w\xi}\right)^n \tag{55}$$

in which η represents the coefficient of kinematic viscosity and n is a number of which Sutton (1953) gives the value $n = 1/4$. The solution of the diffusion problem would be to find a function for the distribution of the concentration, that for a given $R\,(\xi)$—in addition to other conditions—also fulfills equation (54). From equations (54) and (55) it follows that

$$\sigma_x^2 = 2u'^2 \int_0^T \int_0^t \left(\frac{\eta}{\eta + u'^2 \xi} \right)^n d\xi \, dt. \tag{56}$$

In integrating we may neglect terms of the order of η since they are small compared to u'^2T, thus obtaining

$$\sigma_x^2 = \frac{1}{2} C^2 \, (uT)^{2-n} \tag{57}$$

in which C^2 represents the general coefficient of diffusion given by

$$C^2 = \frac{4\eta^n}{(1-n)\,(2-n)u^n} \left(\frac{u'^2}{u^2} \right)^{1-n} \tag{58}$$

For $n = 1$ equation (57) has the form

$$\sigma^2 = 2Kt \tag{59}$$

which corresponds to Einstein's law for Brownian motion.

For turbulent diffusion over an aerodynamically rough surface not only kinematic viscosity η must be taken into account but also macroviscosity, caused by rough ground and called N by Sutton (1953). Since η is surpassed by N many times in size, η can be substituted by N in equation (58) so that the general diffusion coefficients acquire the form

$$C_y^2 = \frac{4N^n}{(1-n)\,(2-n)\,u^n} \left(\frac{u'^2}{u^2} \right)^{1-n} \tag{60}$$

If Q is the strength of the source, i.e. the quantity of diffusing substance given up by the source in time unit, then according to Sutton (1953) the function sought for concentration within the continuous point source is

$$s\,(x, y, z) = \frac{2Q}{\pi C_y \, C_z u x^{2-n}} \times \exp - \left[x^{n-2} \left(\frac{y^2}{C_y^2} + \frac{z^2}{C_z^2} \right) \right]. \tag{61}$$

Since the significance and origin of dimensions C_y and C_z are sufficiently known, we can use this function, derived from the theory of turbulent diffusion, in the observations that follow about the concentration change along the course of flight.

B. Horizontal Concentration Change

A spore cloud, arising from a sufficiently high point source, has the shape of a horizontal cone. The tip lies at the source of the spores, and the base points toward the downwind. A section through this cone, at a right angle to the average wind direction, has the shape of an ellipsis, since the turbulent dispersal of spores is smaller in vertical direction than in the horizontal one. Wilson and Baker (1946) demonstrated this experimentally by using small puffs of ammonium chloride. Brunt (1934) established the ratio of horizontal to vertical components of turbulence as 1.59 : 1 by means of a double wind vane. By catching spores, Wilson and Baker (1946) found a similar ratio 1.55 : 1. The diffusion coefficients C_y and C_z in equation (61) are given by Sutton (1953) as $C_y = 0.4$ and $C_z = 0.2$; thus, they also show the elliptical shape, but with different axes. But, under the conditions of equation (61) Sutton (1953) proved that these values for diffusion coefficients describe the concentration change will distance with sufficient accuracy. Therefore, for simplicity's sake, these values are used.

The form of the spore cloud changes when the ground, as a limiting surface, prevent downward dispersal. In most cases the spore sources lie on the ground or close to it. Therefore we shall consider first the horizontal change in concentration at the ground. The spore source is located at the point $x = y = z = 0$. Consider the change in concentration along the x-axis downwind, i.e., the concentration change with increasing x when $y = z = 0$. Assume a continuous point source with arbitrary but firm values for strength Q and the average wind velocity u. This gives the decrease of concentration in the direction of the wind (along x-axis) based on equation (61) (Fig. 4.19). Since we chose arbitrarily the values of Q, u and x, as long as they are consistent with one another, Fig. 4.19 shows the shape of a curve for the general case.

We can see from Fig. 4.19 that as the distance increases, the concentration rapidly decreases to small values. This explains Kerling's (1949) statement, for example, that even though the infection of peas by *Mycosphaerella pinodes* spreads in the direction of the prevailing wind the strength of attack decreases with distance from the source of inoculum. We also understand the results of studies by Bateman (1947) about pollen dispersal and by Parker-Rhodes (1951), about Basidio-mycetes of Skokholm Island; Parker-Rhodes obtained deposits of pollen or of spores only over short distances. We now understand how Gregory (1952) concluded that 99.9% of spores should fall to the ground within the first 100 meters.

Buller and Lowe (1910) established that the ratio between the number of microorganisms deposited on a horizontal surface and the number of microorganisms found in a certain air volume fluctuates greatly. Durham (1944) came to the same conclusion: there is no close connection between the number of the spores deposited and the actual spore concen-

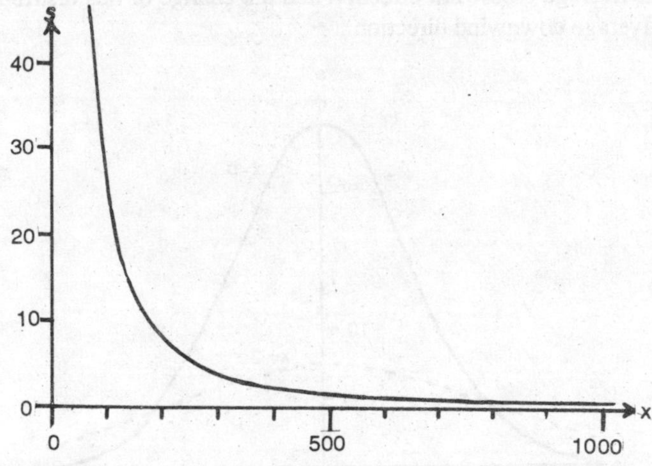

Fig. 4.19. Horizontal change of concentration (s) on the ground with increasing distance (x) from a continuous point source of spores, in the downwind direction.

tration. Just as the spore concentration rapidly decreases to small values with distance, so do the chances of catching spores on a sticky surface decrease. Small wonder that investigators have concluded that the spores precipitate when the effect of turbulent diffusion is not taken into consideration. Figure 4.19 and equation (61), used for calculations, show how untrue this is. In this example and in accordance with hypotheses, no particles have precipitated yet from the spore cloud, but dispersal in a horizontal and a vertical direction have taken place, and turbulent diffusion is responsible for this. This diffusion is actually not constant, even though we have considered it so for simplicity's sake. Reckoning with this we can also understand, at least partly, the results of Bateman (1950), according to which species are not spread in a normal frequency around their source, but the distances are greater and under certain conditions show a digression above the expected value.

Figure 4.20 shows how the concentration changes in a plane normal to the average wind direction, along the y-axis, where $z = 0$, i.e., the concentration is observed on the ground. The change in concentration in the crosswind direction can be observed only at a limited distance from the spore source. Because the downwind concentration according to equation (61) is heavily dependent on distance, two distances were chosen, $x = b$ and $x = 2b$, in order to describe the distribution of the concentration in the average crosswind direction and the change of this distribution in the average downwind direction.

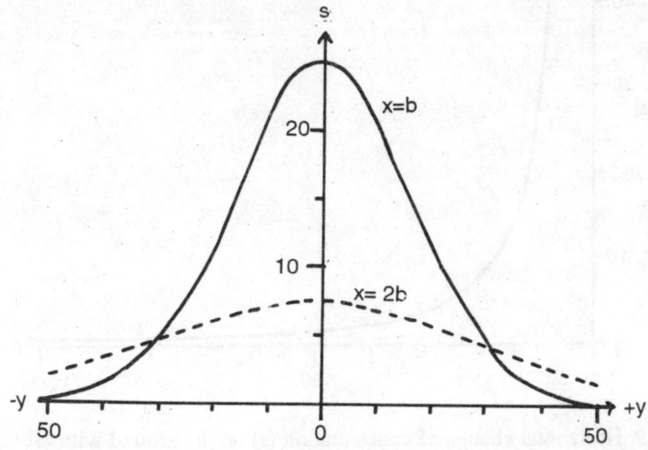

Fig. 4.20. Horizontal change of concentration (s) in the crosswind direction on the ground at two distances, $x = b$ (——) and $x = 2b$ (– – –), from a continuous point source of spores, located at the point $x = y = z = 0$.

As Fig. 4.20 shows, the concentration decreases sharply along the y-axis on either side of the point $y = 0$. If we double the distance along the x-axis where we observe the concentration change and measure it at the value $x = 2b$ instead of $x = b$, then the maximum concentration is much lower, as expected. In the vicinity of $y = 0$ the spore concentration is lower than at a distance of $x = b$. But at a greater distance in the crosswind direction, $y = \pm 50$, the spore concentration is visibly higher than at distance $x = b$. The increasing lateral dissemination through turbulent diffusion leads (at a given distance from point $y = 0$) to the increase of concentration in the periphery at the cost of the concentration in the center; the quantity of dispersed spores remains the same since according to assumption, on precipitation from the spore cloud occurs. The curves offer the picture of a random distribution as we know it from the typical bell shape of Gauss' normal distribution. The difference between the distributions for $x = b$ and $x = 2b$ could be expressed by their different standard deviations. This parameter could be used as a measure for dissemination. Thus, for instance, the decrease in concentration with lateral distance would follow the normal curve and $\pm 3\sigma$ units would include 99.73% of this total. A limited value could be established on the basis of phytopathological considerations, below which the probability of infection is negligible. This measure for "dispersal" could be of extraordinary value for epidemiological questions.

Sutton (1947) pointed out that the particles in a cloud, subject to turbulent diffusion, are distributed normally about the center with a certain standard deviation. This increases steadily when the cloud is expanded into free air through turbulence. He calculated the standard deviation for smoke and gases under various turbulence conditions. Gregory (1945) used Sutton's equation successfully for the question of spore dispersal. He was one of the first in phytopathology to use the modern meteorological presentations about turbulent diffusion for the problem of agent distribution. Empirical formulas for spore dispersal were established several times. For example, Wilson and Baker (1946) described the dispersal of *Sclerotinia laxa* through fruit trees by means of equation (62)

$$y = \frac{A}{x^p} \tag{62}$$

where y is the ratio between the percentage of blossom infection in trees at a distance from the source and the percentage of blossom infection of trees serving as a source. The horizontal distance from the source is x, while A and p are constants that, according to Wilson and Baker (1946), depend on wind velocity and possibly also on other factors. Although Wilson and Baker (1946) recognized that turbulence as well as wind velocity play an important role, their empirical equation (62), with which they could obtain very good results, could not clarify the actual effect of turbulence on dispersal. Gregory's (1945) equations are no longer of purely empirical nature. The dimensions contained in them are defined from physical considerations.

The comparison between theory and observation involved certain difficulties. These arose primarily from our inability to measure spore concentration because of the lack of adequate instruments. Schrödter (1952a) used the Zeiss-Konimeter for measuring the spore content of the air, but the air volume samples taken with this apparatus were too small to yield positive data. On the other hand, the "Cascade Impactor" (May, 1945) made such studies possible. Previous experiments, however, were based mainly on the number of spores adhering to a sticky surface or on the number of infected plants around an infection source. As was shown above, such data could give no information about the actual spore concentration in the air because they depend on too many factors. But since the spore concentration among these factors also represents a significant dimension in unit volume of air, such results can give information, if no quantitatively at least qualitatively, about the seasonal variation in spore content of the air, as established by Horne (1935) and Hyde and Williams (1946). With this sedimentation method Durham (1942) established that a great "spore shower" occurred in October, 1937, over the whole eastern area of the United States, in which the deposit on horizontal surfaces was up to 800–1200 per cm.^{2} a day at several localities.

Observations give little information about the magnitude of Q, nor can a measure of C_y or C_z be derived from them although data on wind velocity u are mostly available. Yet these values are necessary for using equation (61). Stepanov (1935) set free 1.2×10^9 spores of *Tilletia caries* at an altitude of 1 meter so as to catch them at varying distances leeward from the spore source on sticky surfaces. If the data by Lambert (1929) on dispersal aecidiospores of *Puccinia graminis* are observed, or those by Buchanan and Kimmey (1938) on the horizontal dispersal of infection with *Cronartium ribicola*, or by Bonde and Schultz (1943) on infection dispersal by *Phytophthora infestans*, in which case the spore source or the infection focus was exactly known, the same characteristic curve is found as in Fig. 4.19 for horizontal change of spore concentration. Thus it is possible on the basis of certain assumptions to find a direct relationship between theory and the results of such studies.

For this comparison we have chosen the data of Stepanov (1935) and of Wilson and Baker (1946) on the horizontal change in the number of *Lycopodium* spores with distance from the source. To obtain numerical values for use in equation (61) we must assume that the spore number established in these studies are equal to the number of spores per unit volume. We further consider the diffusion coefficients $C_y = 0.4$ and $C_z = 0.2$ as valid also for these experiments. Since strength Q is only known hypothetically through Stepanov's (1935) data (a measure for the spore quantity dispersed per time unit is lacking), we choose the value of Q so that values for concentrations calculated by means of equation (61) for the first distance, data x, is approximately the size of a measured numerical value. This comparison is therefore only approximate.

The result is shown in Fig. 4.21. The dotted curve (*a*) corresponds to Stepanov's data (1935), the dotted curve (*b*) to data by Wilson and Baker (1946). The unbroken curves, on the other hand, represent the calculated values according to equation (61) under the above assumptions. The agreement between theory and observation is considered very good.

Thus equation (61), obtained from the theory of turbulent diffusion, describes the horizontal concentration change in a sufficiently exact manner. The horizontal concentration change follows a logarithmic law. It is directly proportional to the strength of the source and indirectly proportional to the diffusion in horizontal and vertical directions. It decreases with distance, not with x^2 as Wilson and Baker (1946) assumed, but with a somewhat smaller exponent given by $2-n$, According to Sutton (1953) the value $n = 1/4$ can be applied.

C. The Vertical Concentration Change

With the help of equation (61) we can also demonstrate the vertical concentration change just as we can the horizontal concentration change.

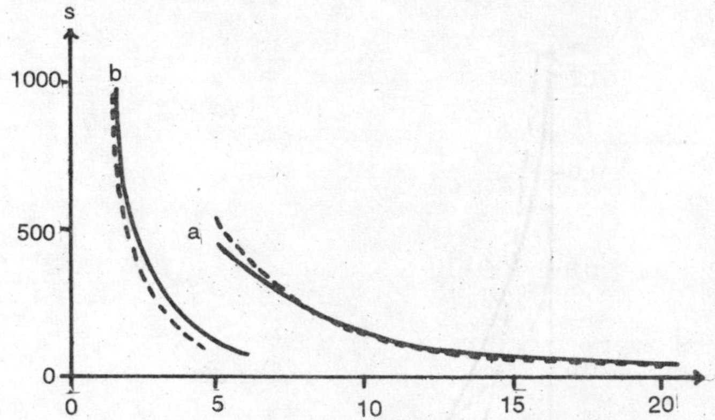

Fig. 4.21. Comparison between observed values (– – –) and values calculated according to equation (61)—(——) of horizontal concentration change. (Observed values (*a*) according to Stepanov, (*b*) according to Wilson and Baker).

Here again the vertical concentration change must be observed at a distance from the spore source. Putting the source at the point $x = y = z = 0$ and observing the concentration change with z at positions $y = 0$ and $x = a$, $x = 2a$ and $x = 3a$ with firm values of Q, u, C_y and C_z, we obtain the curves shown in Fig. 4.22 based on equation (61).

Figure 4.22 shows the decrease in number of spores with height in the expected logarithmic form. In comparing the three curves for distance, $x = a$, $x = 2a$, and $x = 3a$, we recognize that the decrease in number of spores with height becomes less as distance from the spore source is increased, while at the same time the spore concentration rapidly decreases near the ground. Then we obtain the same picture as that already demonstrated in Fig. 4.20 for the horizontal change in spore concentration in the crosswind direction. Close above the earth's surface the concentration decreases as distance increases, but higher above the earth's surface it increases at first.

For the vertical concentration change, too, it is not easy to find experimental data, with the help of which a comparison between theory and observation is possible. The vertical dispersal of spores has usually been studied with airplanes equipped with spore traps. These studies were made in order to investigate how high up spores and pollen could be found. Although no conclusions were drawn as to the change in spore

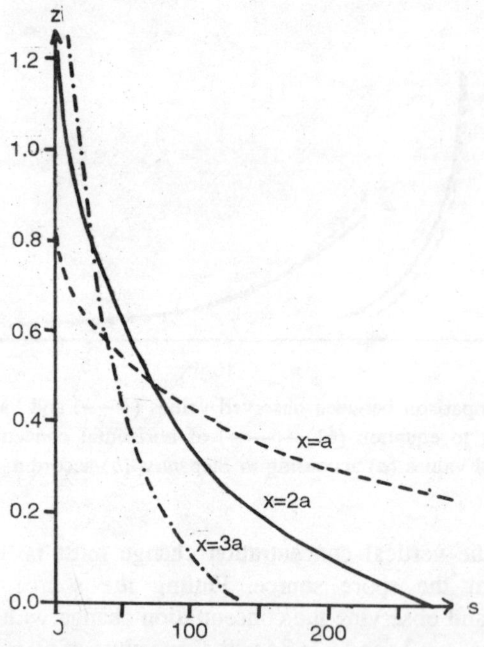

Fig. 4.22. Change of concentration of spores (*s*) with increasing altitude (*z*) at a given distance from a continuous point source of spores.

concentration with altitude within a given spore cloud, the studies did demonstrate a general decrease in number of spores with altitude. These results were later confirmed repeatedly. Thus MacLachlan (1935) used in airplane to catch basidiospores of *Gymno sporangium* at different altitudes over a heavily infected area. Craigie (1945) conducted similar experiments to determine the role of air-borne spores in the dispersal of rust to western Canada. At altitudes between 1000 and 5000 ft. surfaces were exposed by airplane for 5 to 15 minutes and the number of spores per unit area was determined for various altitudes extending as high as 14,000 ft. At 1000 ft. 24,200 spores per unit area were found; at 5000 ft., 7560 spores; at 10,000 ft. 10S spores; and at 14,000 ft., 108 spores. Thus, under these conditions the spore concentration decreases logarithmically with height.

This relation was also demonstrated by Rack (1957) for the air stratum near the ground with simple spore traps set at heights of less than 2.5 meters. Only the relative spore content in the air and its change with height can be established from these studies because of the nature of the instrument used. To use these data to compare theory and observation we must again make plausible assumptions about the unknown parameters. The result of this comparison is given in Fig. 4.23. The agreement between calculated and observed values is good, which indicates that equation (61) describes the vertical change in concentration accurately.

The general validity of this relation between height and spore concentration can also be examined by comparing the curve given in Fig. 4.22 with the results of MacLachlan (1935) and Craigie (1945). Figure 4.24 presents the comparison graphically. Data from which each curve is constructed were converted from the original data to the same relative scale to make them comparable in a graph. This was done by determining the ratio of values of S, the spore concentration, at the lowest altitude in each series and multiplying the original data by this ratio. As Fig. 4.24 shows, the vertical change in concentration calculated by means of equation (61) is qualitatively the same as the observed change in spore concentration with height. Both the horizontal and the vertical change in concentration follow a logarithmic law. It is directly proportional to the strength of the source and inversely proportional to the turbulent diffusion in both the horizontal and vertical directions and is also inversely proportional to the $(2 - n)$ power of the distance from the source of spores.

In the special case when the source of spores does not lie on the earth's surface but occurs at a height h above the ground, z does not equal 0, but instead $z = h$. When the spore source is sufficiently high above the ground and observations are made relatively near (horizontal) to the source, the vertical concentration of spores can be computed by means of

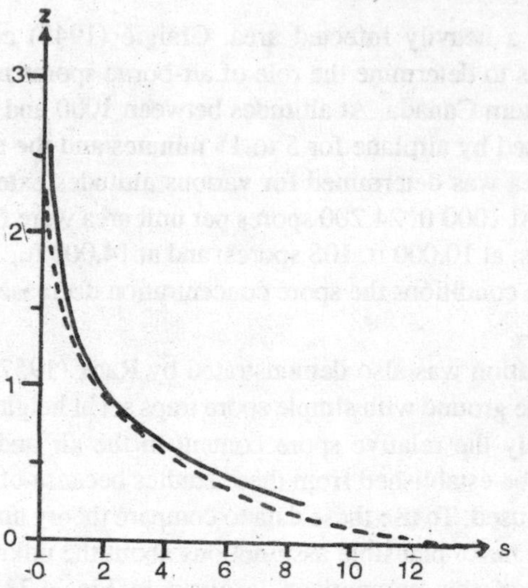

Fig. 4.23. Comparison between calculated (– – –) and observed (——) values of spore concentration (s) with altitude.

equation (61). In this case we chose the coordinate system in such a way that the zero point lies at $x = y = 0$, $z = h$, and altitude z from equation (61) is counted from h, positive upward and negative downward. The question arises whether the diffusion coefficient C depends upon height. Sutton (1953) shows that nature of the relation between the diffusion coefficient C at height z in meters above the ground by the empirical formula

$$C = C_0 - 0.075 \times \log_{10} z \tag{63}$$

in which C_0 represents the diffusion coefficient on the ground. As can be seen, C varies but little with altitude and when small altitudes are involved, we can safely assume that C remains constant. For small altitude differences Sutton (1953) showed that the ground did not act as a limiting surface in vertical distribution of concentration, e.g., the smoke coming from a factory chimney at first spreads as though the earth's surface were not a limiting surface. Thus the vertical concentration change in symmetrical about a line parallel with the earth's surface that goes through the source of spores at height point $z = h$. As Fig. 4.20 shows, such symmetry is to be expected. Studies by Wilson and Baker (1946) indicate that this is true both in theory and in fact (Fig. 4.25).

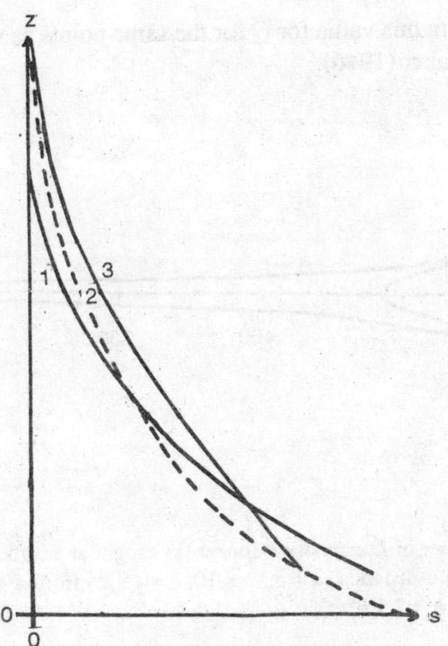

Fig. 4.24. Comparison between observed (——) and calculated (– – –) values of spore concentration (*s*) with altitude. (Curve 1 from observations by Graigie (1945), Curve 3 from observations by MacLachlan (1935), curve 2 from Fig.4.22, where $x = 2a$.

Wilson and Baker (1946) studied the change in spore concentration in the air by freeing *Lycopodium* spores from a source at an altitude of 7.5 ft. and by catching spores on glass plates at various distances from the source. The spore cloud downwind had a sphere-shaped form. Also, as Fig. 4.25 clearly shows, there was a rapid decrease in spore number in the horizontal direction and a nearly normal distribution in the vertical, with the center of the spore cloud as an axis.

A comparison of these observations with theory is supplied by Fig. 4.26, calculated on the basis of equation (61). The value of Q is not obtainable from the data of Wilson and Baker (1946). A value of Q was assumed, such that they estimated value of concentration corresponds to the highest spore number observed by them. This assumption is made to permit a qualitative comparison. All other values of concentration are calculated from this value for Q for the same points as were observed by Wilson and Baker (1946).

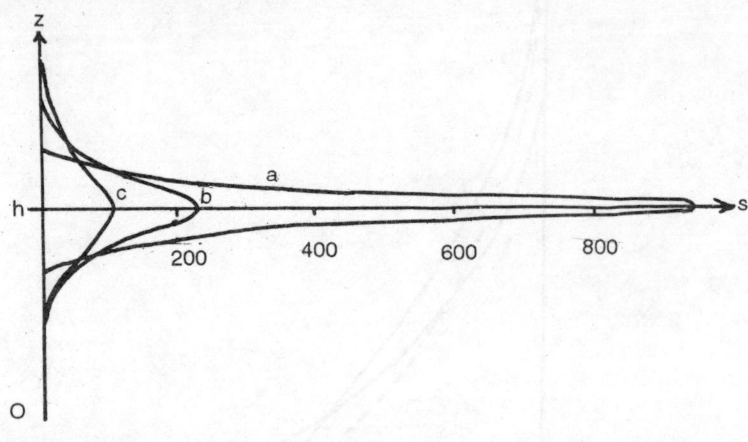

Fig. 4.25. Number of *Lycopodium spores* (s) caught at a different height (z) and at a different distance ($a = 5,\ b = 10,\ c = 15\ ft.$) from a spore source at the altitude of $h = 7.5\ ft.$

Figures 4.25 and 4.26 correspond almost completely. The agreement of observations with theory is unusually good. Thus collecting spores, as Wilson and Baker (1946) and many others have done, can answer questions about spore dispersal when these questions are qualitative rather than quantitative.

D. The Concentration at Ground Level with Elevated Source

Sources of pathogenic spores are mostly on the ground or so near the ground that we hardly have to consider their height. This is especially true because we can regard the upper limit of the closed cover of vegetation as the surface of the earth when vegetation is relatively low. For higher vegetation, e.g., fruit trees, this is not true. From an epidemiological point of view the question of spore concentration at any point is of little interest. The change of concentration with the altitude of the source along a parallel to the x-axis was discussed in the previous sections. We will now give our attention to the concentration of spores near the ground when the spores arise from a high source.

Directly under the source the concentration will be naturally zero. Then the concentration will increase at first. However, since the spore cloud becomes more diffuse with distance because of turbulent diffusion,

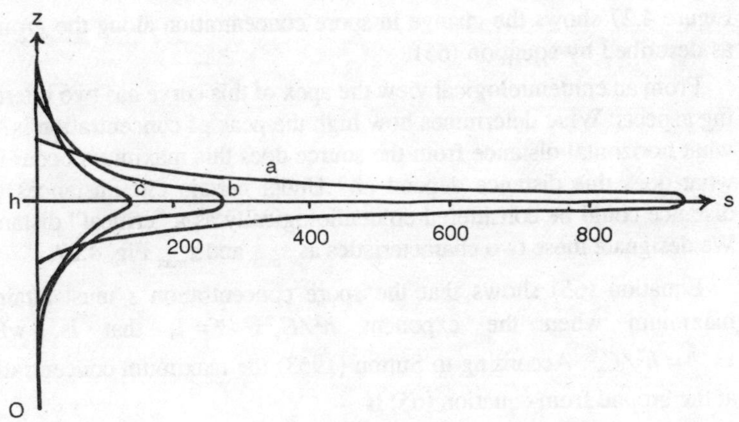

Fig. 4.26. Calculated spore concentration at different altitudes and at different distances from a source of spores at an altitude $z = h$. (Based on data of Wilson and Baker (1946) in Fig.4.25.

the concentration will finally decrease again. Thus, from a high source of spores the shape of the curve relating distance to spore concentration at the ground is not simple.

If the source is at the point $x = y = 0$, $z = h$, then, according to Sutton (1953), the equation for the concentration at one point of the atmosphere with coordinates x, y, z is

$$(s\,x\,y, z) = \frac{Q \times \exp\left(-\dfrac{y^2}{C_y^2\, x^{2-n}}\right)}{\pi C_y C_z u x^{2-n}} \times \left[\exp\left(-\frac{(z-h)^2}{C_z^2\, x^{2-n}}\right) + \exp\left(-\frac{(z+h)^2}{C_z^2 x^{2-n}}\right)\right] \quad (64)$$

The dimensions contained in this equation are all known from previous explanations. When $h = 0$, equation (64) simplifies to equation (61). When we ask only how the concentration changes at the ground, and remembering that the source is at height h above, we can set $y = z = 0$ into equation (64) and simplify it to

$$s\,(x)_{y=z=0} = \frac{2\,Q}{\pi C_y C_x u x^{2-n}} \times \exp\left(-\frac{h^2}{C_z^{\,2} x^{2-n}}\right) \quad (65)$$

Figure 4.27 shows the change in spore concentration along the ground, as described by equation (65).

From an epidemiological view the apex of this curve has two interesting aspects. What determines how high the peak of concentration is? At what horizontal distance from the source does this maximum occur i.e., what does this distance depend on? Under certain circumstances this distance could be considered epidemiologically as a "critical" distance. We designate these two characteristics as s_{max} and x_{max} Fig. 4.27.

Equation (65) shows that the spore concentration s must attain a maximum when the exponent $h^2/C_z^2x^{2-n} = 1$, that is, when $x^{2-n} = h^2/C_z^2$. According to Sutton (1953) the maximum concentration at the ground from equation (65) is

$$s_{max} = \frac{2\,Q}{e\pi u h^w} \times \left(\frac{C_z}{C_y}\right) \tag{66}$$

Thus the maximum concentration at the ground is directly proportional to the strength of the source and inversely proportional to the wind velocity and to the square of the altitude of the source above the ground.

From the condition leading to the derivation of equation (66) from equation (65) we arrive at the distance of the point of maximum concentration from a point under the source

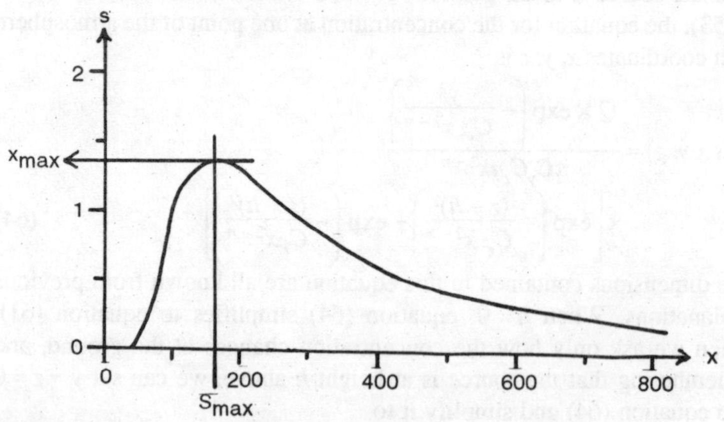

Fig. 4.27. Change of concentration on the ground (s) with increasing distance (x) from a spore source at height (h) above.

$$x_{\max} = \left(\frac{h_2}{C_z^2}\right)^{1/(2-n)} \tag{67}$$

The distance of maximum concentration is approximately proportional to the height of the source, but it is independent of the horizontal wind velocity. This should not be overlooked in evaluating the "critical" distance.

The accuracy of this theoretical result cannot be tested because no observations are available. The studies of Wilson and Baker (1946) involve only relatively short distances, and extrapolations of curves from their data would be unreliable because the number of spores decreases so rapidly as the distance increases. They can serve conditionally, however, to test the accuracy of the theory if we consider the number of spores observed beneath sources at varying height. From their observations we cannot derive the true values for either $s_{\max}$ or $x_{\max}$. However, we can contrast the observed values with curves calculated for this special case from equation (65), and this is done in Fig. 4.28.

The comparison is most alike when spores are counted at a point 2 ft. beneath the source, i.e., $h = 2$. The agreement is also good when $h = 1$. Although there are but three observed values alone, we also know that directly under the source, i.e., at $x = 0$, $z = h - 1$, the number of spores must be zero, thus giving four observed values to which the curve can be fitted.

Thus, when the source is high, the concentration of spores near the ground increases with distance from zero (directly beneath the source) to a maximum at a distance that varies approximately with the height of the source, and then decreases to low values as the distance increases further.

We have seen how the spore concentration decreases to very small values even at a short horizontal distance from the source. Knowing how rapidly spore concentration decreases with distance we can understand why in many spore-trapping experiments, no spores have been detected and also why an astronomical number of spores must be involved in dispersal if successful infections are to occur. Purely biological causes increase the number of spores required. If at a great distance from the source, there is to be a measurable probability for successful infection, the spore output has to be very large indeed. A consideration of turbulent dispersal alone shows that our calculations must involve sources of strength Q, involving numbers of the magnitude of 10^6 to 10^8 spores. Spore sources of this strength have been observed. In local pollen clouds,

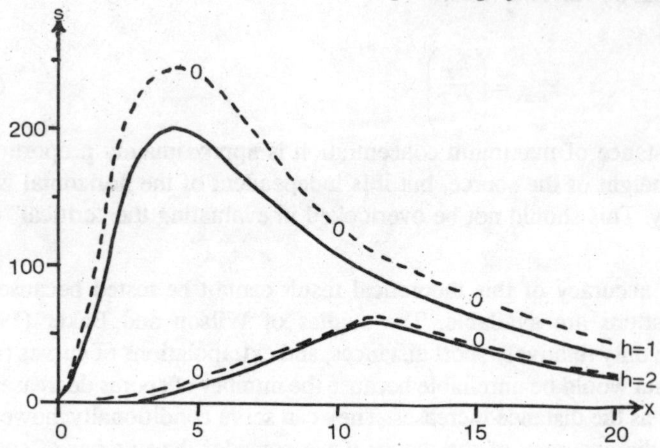

Fig. 4.28. Comparison between calculated (——) and observed (– – –) spore concentrations at ground level for sources of spores of varying altitude.

visible to the naked eye, Durham (1947) measured concentrations from 2×10^6 to 10×10^6 pollen grains per cubic meter of air. According to Buller (1909) reports that *Calvatia gigantea* contained about 7×10^{12} spores. White (1919) reports that *Ganoderma applanatum* can eject 3×10^{10} spores a day for 6 months, while according to Stevens (1911) a single apothecium of *Sclerotinia sclerotiorum* can produce about 3×10^7 spores. According to Ingold (1953) a *Penicillium* colony, with a diameter of 2.5 cm., can develop 4×10^8 conidia; an ear of corn infected by *Tilletia carics* can contain 1.2×10^7 blight spores; the fungus *Daldinia concentrica* can eject more than 10^8 spores a day. Gigantic numbers of spores indeed are needed for dispersal.

IV. LANDING

Duration and range of flight of spores are know from the theory of agent flight. Therefore we known *when* and *where* they land. The discussion was based upon the "statistical mass" of spores and dealt with the center of gravity of the spore cloud. To close the cycle of dispersal, only the velocity with which the spore cloud as a whole gets to the ground can be discussed from the statistical point of view. But the question as to how the single spore lands from the turbulent air is not answered. However, flight and landing are the decisive components for dispersal. Therefore we shall now consider the problem of deposit.

A. The Problem of Deposit

In the following discussion the term "deposit" is used in the broadest sense. It encompasses the deposition on surfaces of spores during transport through the air, whether the surfaces are horizontally or vertically oriented, and whether or not the deposition is induced by external circumstances. The processes of deposition cannot be studied adequately under natural conditions, i.e., outdoors. Moreover, in our considerations, the transition has to be made from the total mass of spores to the behaviour of individuals or groups of individuals.

The whole complex problem of deposition in phytopathology has heretofore been neglected. A few studies have been concerned with the visible result of deposit processes rather than the deposit process itself. Yarwood and Hazen (1942) made studies on how the conidia of *Erysiphe graminis* are oriented during deposition, and thus touched on an important question. Gregory (1952) concerned himself with deposition in a more extensive way and made valuable experimental observations on exactly defined surfaces.

Generally speaking, the very same forces that are responsible for the transportation of the propagules through air are also decisive for deposition of spores. Deposition is but a part of the transfer process. It is the transfer particles from the air to surfaces. Up to now we have observed only the transfer of particles from the surfaces in and through the air. Important forces for deposition are gravitation and fall velocity, under whose influence the spores sediment slowly. Another effective force develops from the horizontal air movement, resulting in the deposition of spores at perpendicular surfaces during their movement in the general current. This process of impaction can deposit spores somewhat forcibly. Turbulence is also an effective force in deposition. Deposition occurs on the upper side of plant organs and on their lower side as well, as a result of motion of particles directed vertically upward as well as downward. Deposition is also expedited by washing out through precipitation. The following section will consider these processes in more detail.

B. The Terminal Velocity

In Section II we considered the motion of the spore cloud and interpreted the definitions of probable flight line, flight range, flight altitude, and flight duration that were introduced by Rombakis (1947). Since the "where" and "when" have been treated within the motion complex, we can readily clarify the problem of terminal velocity. To this end, the data by Rombakis (1947) are again helpful. By referring to the derivations given in Section II we can readily obtain a precise definition for end velocity. The "probable end velocity" is the end velocity of the "center of gravity" of the spore cloud.

Turning again to some equations from Section II we have equation (32), from which we obtained the equation of probable flight line

$$z = 0.4769\sqrt{4at} - ct \tag{32}$$

and obtained the equation of the maximum probable flight altitude, with the help of equation (32), from the condition

$$\frac{dz}{dt} = 0.4769\sqrt{4a}\,\frac{1}{2\sqrt{t}} - c = 0 \tag{44}$$

This equation says merely that the vertical components of velocity at the highest point of the flight line are equal to zero.

If we ask about the end velocity, with which the spore cloud arrives at the ground during the landing process, we naturally refer to a vertically directed component of velocity. Thus, starting from equation (44), we have only to ask about the velocity dz/dt at the end of the time of flight, that is, the time $t = \tau$. The flight duration, τ, is given by

$$\tau = 0.91\,\frac{A}{\delta c^2} \tag{51}$$

which was derived from equation

$$\tau = (0.4769)^2\,\frac{4a}{c^2}. \tag{50}$$

If we take equation (44) at the time $t = \tau$ we get

$$\left(\frac{dz}{dt}\right)_\tau = 0.4769\sqrt{4a}\,\frac{1}{2\sqrt{\tau}} - c. \tag{68}$$

From equations (50) and (68) equation

$$\left(\frac{dz}{dt}\right)_\tau = 0.4769\sqrt{4a}\,\frac{c}{2\times 0.4769\sqrt{4a}} - c \tag{69}$$

follows, from which

$$\left(\frac{dz}{dt}\right)_\tau = -\frac{1}{2}c \tag{70}$$

results. This is the "probable end velocity." Thus the spores fall to the ground with a velocity that is independent of all transfer force except gravitation and is just half as great as the fall velocity due to gravitation in calm air. Since this fall velocity is generally very small, especially as compared with the horizontal wind velocity, we must establish that the landing of the spore cloud is nothing but a slow sedimentation under the influence of gravitation, independent of turbulence and horizontal wind velocity.

C. Sedimentation and Impaction

The conclusion that the landing of a statistical mass of spores occurs by sedimentation cannot be applied to the individual spore, except in calm or laminally streaming air, when all vertically acting forces are eliminated. On clear nights free of wind, when the air near the ground has cooled down by radiation, the laminar or "quasilaminar" layer can rise up to a few meters Under these conditions the air soon becomes free of spores through sedimentation, as Tyndall (1881) has shown. In free air, where vertically oriented forces of turbulence influence the spores again and again, sedimentation cannot play a very great role as a factor in landing.

Normally during the day the laminar layer exists only as boundary layer directly at the surfaces where at the most it gains a few millimeters in thickness. Since the boundary between the laminar surface layer and the turbulent air current is not constant, spore-contains whirls can penetrate into the laminar layer, as described by Gregory (1952), and leave behind small volumes of air containing spores in an interchange with air free of spores. Spores thus brought into the laminar layer settle then under the influence of gravitation and deposit before new whirls from the turbulent layer can reach them and take them away. In this case forces other than sedimentation play a primary role.

When the wind brings to the surface of the plant a volume of air containing a certain quantity of spores, the number of spores deposited is difficult to determine. We must first observe the processes on geometrically simple, exactly defined surfaces. Wind tunnel experiments as carried out by Gregory (1951) seem to be especially adapted for this.

Under different wind velocities Gregory (1951) examined the deposition of spores of *Lycopodium clavatum* on cylinders with diameters ranging from 0.018 to 2.0 cm. Deposition increases with increasing wind velocity and with decreasing cylinder diameter. The effect results mainly from "impaction," a principle described by Sell (1931) for liquid drops. If an object (cylinder, plate, etc.) is set in the path of a spore moving with an air current, the spore moves by its own inertia a bit farther toward the object even when the air has already avoided the obstacle. The length of this "course of inertia" determines whether the particle hits the obstacle and is deposited on it, or whether it is deflected around it without deposition. The length of this course of inertia depends on the size of the particle, the velocity of the current, and the form and size of the object placed in the course of flight. These factors at the same time determine to what degree an object is suited for spore deposit. From aerodynamic observations Sell (1931) shows that in the case of cylinders the impaction efficiency is determined by the dimension-free size

$$k = \frac{v_s v_o}{rg} \tag{71}$$

whereby v_s means fall velocity of the particles in calm air, v_o the current velocity, r the radius of the cylinder, and g the gravitation acceleration. The connection between size k and impaction efficiency E was determined empirically by Sell (1931). Glauert (1946), who considered a similar problem, determined a coefficient that is practically identical with the one by Sell (1931). The values of impaction efficiency observed experimentally by Gregory (1951) in the wind tunnel are lower than would be predicted by the equations of Sell (1931) and Glauert (1946). But the values correspond well to those by Langmuir and Blodgett (1949). In their theoretical treatment of the problem, a set of curves is established for the relation between E and k, which is given from the dimension free size

$$\Phi = \frac{R^2}{k} \tag{72}$$

in which R represents "Reynold's number," already described in equation (11). For values $\Phi = 10$ to $\Phi = 10^4$ the values of impaction efficiency determined experimentally by Gregory (1951) fall within the range of these curves (except for a few extreme values of k).

Deposition by impaction, which results from the translation energy of the particles headed up by the wind, plays the greatest role in the landing of single spores. Gregory (1951) showed that the effect is small when small spores approach large obstacles in the presence of small wind velocities. Large spores, on the contrary, are predestined for deposition by impaction, especially on small obstacles at high wind velocity. Many leaf-disease-causing agents dispersed by wind have relatively large spores (*Phytophthora, Helminthosporium*, etc.) while species that inhabit the ground (*Penicillium, Aspergillus*) are mostly characterized by small spores which are not suited for deposition by impaction.

A high impact efficiency is naturally unsuited to the dispersal of a disease within a dense plant population. According to Johnstone et. al.

(1949), the capacity of a particle to penetrate into a closed vegetation covering is inversely proportional to the impaction efficiency of vegetation, since a high degree of efficiency greatly limits the mobility of spores. According to Gregory (1952) a spore size of 10μ, the most frequent diameter of spores dispersed by wind (e.g., of many ascospores and basi-diospores), represents a good compromise between the originally opposed requirements for dispersal and deposition. The large-spored leaf and stem parasites are quite correctly labelled by Gregory (1952) as "impactors," while the small-spored ones seem to be special "penetrators" that are deposited rather by sedimentation.

The usual vertical sticky deposits, used in routine catching of spores and pollens have only a very small impaction efficiency. The results so obtained, therefore, offer an incomplete picture of dispersal of pathogens. This accounts for the frequent divergence of data on dispersal and for the discrepancies between theory and observation.

According to Gregory (1952) "turbulent deposit" is another factor in the landing of single spores. This effect was noted by Durham (1944) while catching pollen, and he found that the deposit on horizontal surfaces is greater (as would happen from sedimentation) but that a deposit also occurs at the lower side of horizontal surfaces, an effect that cannot be explained by sedimentation. Turbulence transports the spores not only from above to below but also from below to above. Under these forces a deposition occurs on the lower side of surfaces. The turbulent deposit is but a deposit through impaction, at which the particles gain their kinetic energy from the vertical rather than from the horizontal component of the turbulent current. Consequently, only two effects are generally important for the landing of a single spore: sedimentation under the influence of gravitation and impaction under the influence of the movement of the current.

The combined effect of these processes leads to a decrease in spore content of the air. At least in large spores in dry weather there is seldom a consistent increase of spore content in the air. Gregory (1952) concluded that the daily spore ejection of larger spores is in equilibrium with the daily deposit. Meetham (1950) obtained a similar result for smoke impurities in the air, so that the above conclusion seems completely justified. With regard to the theory of agent flight it is questionable, however whether the assumption of a daily equilibrium between expulsion and deposit can be justified in small (or in the smallest spores).

D. The Humidity and Precipitation Effect

In the theoretical discussion we did not take into consideration that through the influence of external events, a "forcible" limitation of spore dispersal is possible. We supposed the falling velocity of particles to be a constant process. In connection with the problem of deposit we are now concerned with the problem of what the consequences are when these assumptions are not fulfilled. Let us, for instance, observe the fall velocity: clearly it actually cannot be constant. A spore that gradually dries out during flight becomes lighter; a spore that swells in humid air becomes heavier. In condensation the accumulation of water by the transferred particles makes them heavier, and in that way, too, their fall velocity changes quite considerably under certain conditions. Since the fall velocity, on the other hand, exerts a great influence on dispersal, the humidity of air tends to limit the theoretically possible dispersal and result in premature deposit. Thus the usual simple methods of bacteriological air studies (exposing of agar plates) produce different results depending on the relative humidity of air. Increasing air humidity results in a decrease in spore concentration. According to Silhavy (1954) the number of germs are generally low (= strong deposit) at a low temperature with high air humidity; at a high temperature with small relative air humidity, on the other hand, they are generally high (= scanty deposit).

It is not easy to find a connection between spore deposit and relative air humidity, since the number of spores caught depends on other factors as well. In his studies of the epidemiology of *Phytophthora*, Raeuber (1957) found a correlation of $r = +0.54$ between the number of spores trapped and the relative humidity. We must assume, accordingly, that the relative air humidity plays an important role in the process of deposition and for the total complex of dispersal. Although studies about this is phytopathology are rare, the assumption seems justified since it corresponds to the results of aerosol research. Studies by Meetham (1954) dealt with the deposit of smoke particles in the presence of fog. Little indeed is known about the mechanism of this type of deposition.

The influence of rain on dispersal and deposit is better known. In single instances rain alone can transport a pathogen from one plant part to another. Aside from this, rain means the end of spore dissemination. For small and very small spores that can barely be deposited through other processes, washout of spores by rain represents one of the few possibilities to get to the ground again after a short period of time. In spore-trapping experiments Rack (1957) showed that the relative spore density (number per surface unit) increases with the increasing total precipitation. The duration of precipitation is also decisively significant for the decrease of particle content in the air, as Raynton (1956) could establish. The rapid filtration of radio-active aerosol particles from the atmosphere by precipitation was described by Mügge and Jakobi (1955), while Herman and Gorham (1957) came to the conclusion from studies of mineral constituents in snow and rain that more substances are washed out of the atmosphere with rain than with snow.

The mechanism of spore washout by rain is closely dependent on raindrop size. Raindrops have a diameter up to 5–6 mm. According to Best (1950) their fall velocity lies between 2 and 9 meters per second. The problem of rain washing of particles from the air was studied by Langmuir (1948), whose theory of collection of particles by raindrops is restricted primarily to sphere-shaped particles. On the basis of this, Greenfield (1957) treated the washout of radioactive particles from the atmosphere and showed that the exchange effectiveness between raindrops and particles depends on the particle size. The number of particles washed out is a function of the quantity and duration of precipitation. On the basis of Langmuir's (1948) theory Gregory (1952) concluded that small spores of *Lycoperdon* and the ground-inhabiting Penicillia cannot be collected by drops with a diameter under 1 mm. In drops with a 2 mm. diameter the effectiveness of collection increases to a maximum of 15% but decreases again with larger drops. For basidiospores of *Psalliota campestris* the collection commences with a drop diameter of 0.2 mm., achieves a maximum of 30% with a drops diameter of 2 mm., and decreases again with further increasing drop size. In *Tilletia caries*, in uredospores of *Puccinia*, and in conidia of *Erysiphe graminis* the collection with all possible drop diameters is larger than zero and achieves a maximum of 80% with a drop diameter of 2.8 mm. The maximum collection is achieved with drop sizes of 2 mm. It is about 25% in small spores with a 4μ diameter and 80–90% in large spores (20–30 μ diameter).

Whether any other factors influence the strength of rain washout of spores is not known. The effect of wettability on washout is unknown. For example, Gregory (1952) notes that spores of *Ustilago perennans* are moistened by water only with difficulty but are very easily collected by rain. When they have reached the ground, they remain, as Burges (1950) showed, in the upper ground layers like other unmoistened spores, and they are not washed into the ground.

The influence of washout by rain on the deposition of spores can be shown by an example given by Gregory (1952): 2 mm. rain elicited a deposition of spores of *Ustilago* about 200 times as great in 2 hours as would otherwise be deposited during a whole day on equal surfaces exposed to wind but protected from rain.

The rain washout means an abrupt end to their dispersal for the majority of pathogens. It strongly restricts the dispersal under certain conditions, especially for leaf and stem pathogens; the ground-inhabiting organisms, on the other hand, have very small spores and but a small chance of returning to the ground in a short time in the absence of precipitation. For them washout by rain is presumably the normal way of landing.

V. VARIOUS PROBLEMS

The circle of questions that are connected with flight and landing of agents is now closed. Certain problems remain that seem important.

A. Particle Stir-Up from the Ground Surface

In the treatment of spore concentration of the atmosphere we assumed that spores arise from a point source, that the spore ejection is continuous and the fall velocity negligible. This picture is somewhat incomplete even if it leads to a satisfactory theoretical interpretation of the processes. Under certain conditions inoculum accumulates in dust consisting of very fine particles of plant residues that are whirled by the wind into the air. Once it gets into the air, this material can be considered a part of atmospheric dust. Just as it does with dust the wind will also stir up and transport rust spores from an infected field. How does the spore concentration of the air change in such a case? Clearly the concentration must increase from the leeward side to the windward side of such a field, since new spores always are added to those that were first stirred up and carried on. It is clear also that this increase in concentration can last only as long as the field is traversed by air currents, and then a concentration decrease should follow, corresponding to the theory already presented on horizontal concentration change. However, it is also evident that the concentration cannot increase in an unlimited manner even when the spore "activity" is ever so large, because after a certain time a type of equilibrium will be reached. Fortak (1957) treated this problem in a theory of dust transfer over a dust-active earth surface. Since his results are pertinent to the problem of spore dispersal, we will discuss the applicable aspects here.

If a wind is blowing with velocity u in the positive direction of the x-axis, the fall velocity of a certain particle is s, the exchange coefficient is η (measured in cm.^{2}/sec.) and the number of particles in volume unit is σ, then the differential equation of the vertical mass exchange is

$$\frac{\partial\sigma}{\partial t} + U\frac{\partial\sigma}{\partial x} - s\frac{\partial\sigma}{\partial z} = \eta\frac{\partial^2\sigma}{\partial z^2} \tag{73}$$

[Compare this with equation (24) in Section II.] For various conditions Fortak (1957a) gave solution of this differential equation that are significant for the theory of dissemination in the stirring up from ground or plant surface. Since only particles having a certain fall velocity with constant wind and constant mass exchange are observed under these conditions, we wish to know their concentration.

For simplification Fortak (1957a) introduced new, independent variables such as $T' = s^2t/4\eta$, $X = s^2x/4\eta U$, and $Z = sz/2\eta$, whose connection with equations (51), (43), and (47) is readily evident in Section II. Equation (73) is simplified with these normal variables to

$$\frac{\partial\sigma}{\partial T} + \frac{\partial\sigma}{\partial X} - 2\frac{\partial\sigma}{\partial Z} = \frac{\partial^2\sigma}{\partial Z^2} \tag{74}$$

On the basis of this solution of the particle transfer for each particular case can be described. This is true not only under the above mentioned concentration increase until a sedimentation equilibrium is obtained, but also throughout the entire process of particle filtration, e.g., by forest or hedgerows.

How the course of spore concentration takes shape on the ground when the wind blow over a "spore active" strip, such as a rust-infected corn field, and how it carries along stirred-up spores over an adjoining noninfected area can be treated theoretically.

According to Fortak (1957a) for the special case when $x < d$, d being the width of the infected strip, the solution is

$$\sigma = \frac{C}{s} \times \psi_{0\,(X,\,Z)} \tag{75}$$

an equation that describes the change of spores concentration in the direction of the wind over the infected strip. For all $x \quad d$, or with the above normal variables written $X \quad s^2d/4\eta U = D$, thus for the area beyond the infected strip in the direction of the wind, the result is

$$\sigma = \frac{C}{s} \times [\psi_{0(X,Z)} - \psi_f 0\,(x - D, Z) \tag{76}$$

and for the spore concentration at the surface of the previously spore-free area, we have, when $Z = 0$,

$$\sigma = \frac{C}{s}\,[4_i^2 \text{ erfe} \sqrt{X - D} - 4_i^2 \text{ erfe} \sqrt{X}] \tag{77}$$

in which functions of the type of erfc x or ierfc x are known functions from the theory of heat conduction. The factor C represents a measure for the dimension of spore stir-up or spore filtration.

Figure 4.29 represents the solution of equations (75) and (77) for a certain width of the infected strip. Since the normal variables were again used here, this figure represents regularity in a general form. For individual conditions of wind and turbulence and for a given spore shape or sedimentation type, the change in concentration can be obtained. It is only necessary to transfer the normal variables X and D into the actual distance coordinates x and d. Thus, for instance, in a strip with a width of 240 meters and under normal conditions of wind and turbulence, the relative spore concentration near the ground will take the course shown

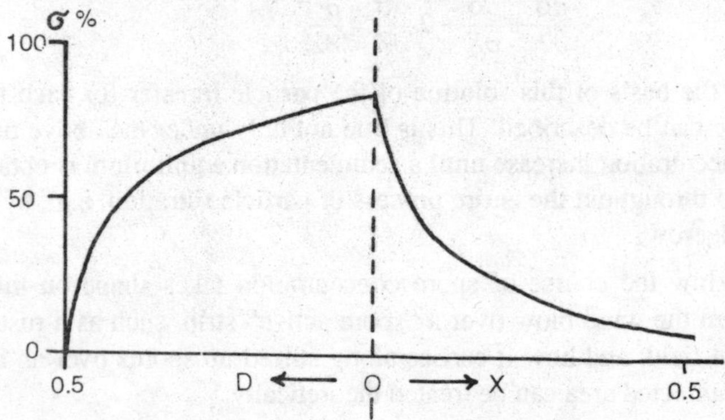

Fig. 4.29. Change of relative spore concentration σ in wind direction over a "spore active" strip of width $D = s^2d/4\eta U = 0.5$ and over an adjoining, originally spore-free area ($X > 0$).

in Fig. 4.30 namely, according to curve *a* with a fall velocity of spore of 1 cm./sec. and according to curve *b* with a fall velocity of 0.45 cm./sec. In small (lighter) spores the sedimentation equilibrium is achieved later, so that at the end of the strip the relative concentration is smaller than the case of larger (heavier) spores. On the other hand, the relative concentration decreases more slowly than the distance from a "spore-active" strip. Thus, almost behind the strip the concentration is higher than for larger spores, and a relative concentration of 10% is achieved only at a distance 50% greater from the strip. If a relative concentration of 10% is considered as the lower limiting value for danger

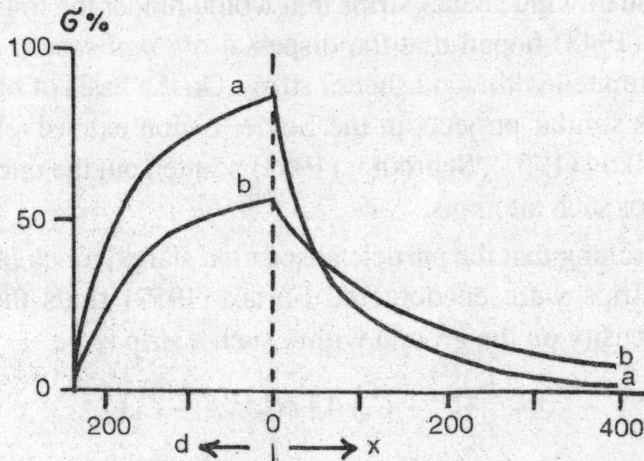

Fig. 4.30. Change of relative spore concentration σ in wind direction over a "spore active" strip of width $d = 240$ m. over an adjoining, originally spore-free area with various fall velocities ($a = 1$ cm./sec., $b = 0.45$ cm./sec.) demonstrated with real distance co-ordinates for normal external circumstances.

of infection, the decrease of fall velocity to about half would increase the danger of infection by 50%. It is shown here too how great is the significance of fall velocity for the dispersal of pathogens.

If we consider that the starting concentration of small spores amounts to many times more than that of large spores, then the differences become even clearer. If, for example, according to conditions underlying Fig. 4.30 the initial concentration of small spores is twice that of the large ones, the number of small spores per volume unit at a 250 meters distance from the spore-active strip would be three times as great as that of large spores; in case of increase in initial concentration by a tenth power, it would be sixteen times as high as that of large spores. In addition to wind and turbulence conditions, the initial concentration and the fall velocity establish either the size of the area in which there is a given danger of infection or the magnitude of the danger of infection within a certain area.

B. Effect of Filtration of Obstacles (Wind Shelter Strips)

The theory of Fortak (1957) allows description of particle filtration by wind from strips. Rows of trees or hedges have been employed as protective screens in phytopathology. Thus, according to Klemm (1938) attempts were made in Rumania to prevent the dispersal of corn rust by installation of wind shelter strips that would hinder the transfer of spores. Ruggieri (1948) hoped that the dispersal of "*mal secco*" of citrus trees could be limited with wind shelter strips. On the basis of his own studies, as well as similar projects in the Soviet Union extensively reported by Grebenšcikov (1951), Schrödter (1952) pointed out the uncertainty of the outcome of such attempts.

By assuming that the particle absorption started much before the wind defense strips were encountered Fortak (1957) finds the solution for particle density on the ground within such a strip to be

$$\sigma = \frac{1}{s}\,[C_1 + C_2)\,4_i^2 \operatorname{erfc}\sqrt{X} - C_2] \tag{78}$$

and for the particle density on the ground behind the wind shelter strip (at the leeward side of the strip) to be

$$\sigma = \frac{1}{s}\,[C_1 + C_2)\,4_i^2 \operatorname{erfc}\sqrt{X} - C_2 4_i^2 \operatorname{erfc}\sqrt{X-B}\,] \tag{79}$$

C_1 is a measure for the particle stir-up, C_2, on the other hand, a measure for filtration, while the normal variable $B = s^2 b/4\eta U$ contains the real width b of the wind shelter strip. Hence, the course of the particle concentration on the ground shown in Fig. 4.31 follows for a given strip width B with diverse filter effectiveness C_2. Comparison with concentration change outside of the wind shelter strip—and uninfluenced by it (dotted curve in Fig. 4.31)—shows clearly how quickly the concentration decreases within the wind shelter. However, the repeated increase of concentration directly behind the defense strip is especially interesting. Thir increase is observed also in practice as shown by Schrödter's (1952a) spore traps and by Illner's (1957) observations about the question of wind shelter and weed dispersal. Both studies show that the usual shelter strips are scarcely to hinder the dispersal of spores or weed seeds.

The question is naturally asked how wide a wind shelter strip has to be if the total content of particles with sedimentation constant s is filtered out. Based on the theory by Fortak (1957) width B' of the "total" filtration can be found with the help of equation (78), which can be written under these conditions as

$$0 = (C_1 + C_2)\, 4_i^2 \operatorname{erfc} \sqrt{B'} - C_2. \tag{80}$$

Solve according to *B'* the result is then the searched-for width of the wind defense strip to

$$B' = \left(4_i^2 \operatorname{erfc}^{-1} \times \frac{C_1}{C_1 + C_2}\right)^2. \tag{81}$$

Since *B* or *B'* are normal variables, known to have been introduced to make the description independent of the wind and turbulence conditions, we have to take into consideration that the real strip width *b* necessary for complete filtration can be completely different, depending on external circumstances.

Thus, theory shows that the installation of wind shelter strips to avoid spore transfers is completely problematical, which explains the various

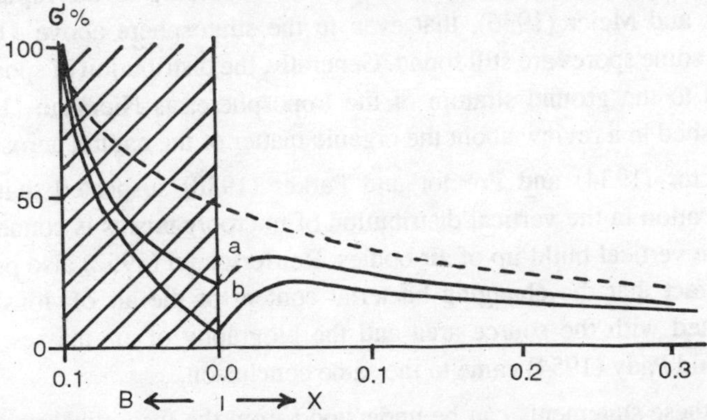

Fig. 4.31. Change of relative spore concentration on the ground in and behind wind shelter strips of width *B* and different filter effect C_2 (*a*: $C_2 = 0.5 \cdot C_1$; *b*: $C_2 = 0.8 \cdot C_1$; – – – outside of the wind shelter).

failures of such attempts. Future studies about the question of wind shelter and spore dispersal will have to pay more attention to the physical side of the problem. For this, a foundation is the theory of dust transfer by Fortak (1957) in its application to the questions of dispersal of disease-producing agents.

C. Spores in Upper Air

The presence of spores in higher atmospheric strata was noted in connection with the problem of epidemiology of corn rust. From the experimental point of view Stakman et. al. (1923) were the first to prove that the forces of turbulence and convection are capable of transporting even the relatively large uredospores of rusts to great altitudes of the atmosphere. Numerous similar studies confirmed this. That only very few spore are caught at these altitudes is insignificant. The fact that spores can be transported to these altitudes at all is quite important from an epidemiological point of view.

In general., at 4000–5000 meters altitude with a time of exposure of 5–15 minutes, a result of 2–3 spores per cm.^{2} can be expected. The continental "steam cap" reaches nearly up to this altitude, as the studies by Junge (1951) about vertical distribution of aerosols over Europe show, in normal exchange and in undisturbed weather. This does not mean that no spores can be found at greater altitudes. The ascent of the American stratosphere balloon "Explorer II" proved, according to the report by Rogers and Meier (1936), that even in the stratosphere above 11,000 meters some spores are still found. Generally, the distribution of spores is limited to the ground stratum of the troposphere, as Niemann (1955) established in a review about the organic matter of the natural aerosol.

Proctor (1934) and Proctor and Parker (1938) suspected that the stratification in the vertical distribution of microorganisms is connected with the vertical build-up of air bodies. Reifferscheid (1952) also points to the fact that the changing bacterial content of the air obviously is connected with the source area and the biography of air masses, and Kelly and Pady (1954) came to the same conclusion.

All these statements can be understood from the theoretical point of view. As far as the flight altitude is concerned, we only have to refer to the explanations in Section II. As Table 4.5 showed, a flight altitude of more than 5000 meters lies completely within the possible range. One has to consider that only the "maximum probable flight height" is concerned here and that according to the definition spores can also be found with equal probability above this point. The flight altitude depends not only on turbulence conditions but also on fall velocity. Hence, very small spores could theoretically achieve a maximum flight altitude of 15.750 meters according to value calculated by Schrödter (1954) in a mass exchange of 50 gm./cm. sec.

We therefore have to establish that the presence of spores in higher atmospheric strata is not unusual but is to be expected from the theory of agent transfer.

D. Intercontinental Dispersal Through the Air

The presence of spores in the high atmosphere is in itself of small interest from a phytopathological point of view. Its significance lies in the fact that it leads to a consideration of whether an intercontinental dispersal of pathogens is possible with air current. Ingold (1953), for instance, asks whether spores can be transferred from America to Europe or vice versa. If we want to consider the problem of intercontinental dispersal theoretically, then we must exclude the question of viability. Whether transfer of spores is possible from one continent to another can be answered in the affirmative, and then even a water surface such as the Atlantic Ocean does not represent an obstacle.

Flight ranges of more than 10,000 km. are quite conceivable (compare with Table 4.2. This is not a purely theoretical result. Mügge and Jakobi (1955) point to the fact that for a broad dispersal of aerosols the so-called "jet-stream" is decisive: a strong west current with which at an altitude of 8–10 km., velocities of 200–400 km./hour are reached. By this "jet-stream," aerosol particles can be transferred in a very short time over very great distances, since through the meandering of this current, rapid transfers into a north-south direction can occur. These and similar currents are probably able to transport spores and pollens (that according to theory and experience are present at these altitudes) over very large distances, even from continent to continent. This assumption is corroborated by various studies and observations. The work of Junge (1951) shows, for instance, that aerosols of extra-European origin occur above the so-called continental moisture cap. Craigie (1945) could make believable the fact that uredospores of rust, found at various points in Canada, must have originated from sources lying very far south, because at the time of trapping these spores, the most northern limit of the area with rust infection was found about 500 miles south of Canadian trapping places. Bergeron (1944) reported about a probable wind transfer of spore and pollens from the area of the lower Yenisey up to southwest Finland. That the ocean does not represent an obstacle can be seen from the fact that according to Niemann (1955) pollen precipitation can be found far up over the ocean. Naturally, spores that have to traverse the ocean can be found only in the highest air strata above it. Spore trappings from the ship's mast, as carried out by Zobell (1942), indicated that the air over the ocean became free of fungal spores. Such results are only evidence of limited strength. Pady (1954) can, on the basis of two transatlantic flights, show a completely different result, which confirms an intercontinental transportation. Accordingly, air masses of polar origin above the ocean can contain about 4 fungal spores per cubic foot, while in air masses of tropical origin, even about 15 fungal spores per cubic foot were found.

All these observations prove that the supposition is justified, not only from this theory, but also from practical experience, that through air currents an intercontinental transfer of spores, even above the ocean, is absolutely possible and probable. Whether this knowledge has a greater practical significance seems doubtful. Even if the propagules retained their viability very such a long period of time, many other conditions would still have to be fulfilled before the intercontinental transfer of spores becomes significant. Through the widening of world trade with plant and plant products man himself has obviously become a major factor in the intercontinental dispersal of plant disease, more effective at any rate than the intercontinental transfer of disease-producing agents through air currents could ever be. This knowledge in no way lessens the great significance that wind and turbulence have for dispersal. This significance does not lie in the theoretically given possibility of overcoming very great distances, but rather in the fact that these factors also determine decisively the dispersal in a limited area. The knowledge of the basic principles of agent dispersal, as they have been demonstrated here, is undoubtedly an important contribution to the understanding of those processes that lead to an occurrence of illness. It herewith establishes the most important foundations of epidemiology.

5

DISPERSAL BY ANIMALS

Some pathogens are carried externally by animals, others internally; some are carried passively, others appear to have an active biological association with their vectors, as the animal carriers are called. Much is known about the spread of several pathogens by man and by insects, but very little about those dispersed by other animals. Most of this chapter is devoted to the spread of viruses, for although there are some diseases of great economic importance caused by animal-dispersed bacteria and fungi, relatively few are plant pathogens.

Knowledge of the methods by which pathogens are dispersed is usually essential before effective control measures can be devised. Two types of spread must be considered: (1) the introduction of the pathogen into a healthy crop, which event may occur over considerable distances, and (2) the spread from plant to plant within a crop.

But let us examine first the pathogens to find out how they are adapted to transmission by animals. Viruses are by definition obligate parasites, and few survive for long outside living cells. Most lose infectivity in extracted plant sap within hours or days, but some, like tobacco mosaic and potato X viruses, can persist on clothing or in plant debris for many months, and are so infectious that they can be carried externally on animals, clothing, or machinery. This is exceptional behavior, and most viruses are transmitted only when an animal, usually an arthropod, feeds first on an infected plant and then on a healthy one. The relationship between arthropod and virus is often close, as we shall see, although it has involved no obvious changes in the animal's morphology.

Bacteria, like viruses, are unable to penetrate plant cuticles and depend for entrance on natural openings or injuries. They depend less on animals for transport, as most are splashed in rain drops, are carried on or in seeds, or in soil, or occasionally are blown by wind. Some have several forms of transport, including animals, but at least one, the cause of cucumber wilt, appears to be entirely dependent on insects. Some bacteria occur in flowers and are spread by pollinating insects; some are contained in slime that oozes from lesions, and insects become contaminated with these as they move about the plant or they may even be attracted to the slime. Others, like some viruses, are transmitted by feeding insects, and some are retained within the animal's body for a considerable period. Most of these do not form resting spores or similar structures, and must survive periods when plants are absent, or weather is unsuitable for growth in diseased plant tissues, or more rarely, in soil or insect vectors. Unlike the fungi, bacteria are not adapted for wind dissemination, nor have they developed really efficient methods of insect transmission as the viruses have.

Most fungal conidia are of the dry spore type and are blown away by the wind, but some have slime spores carried in a sticky liquid, which is dispersed by water or animals, usually insects. These are mostly microfungi, and only a few are pathogenic to plants. This subject was dealt with exhaustively by Leach (1940) who drew attention to the close resemblance in many respects between insect dispersal of pathogens and insect pollination. Insects that both protect a pathogen and provide a means for it to enter a plant obviously provide an effective and prudent method of spread. There are some examples of apparent symbiosis because the insect appears to benefit from its association with the fungus.

I. DISPERSAL BY MAN

Vegetative propagation from infected plants is an all too efficient means of increasing the number, and often the geographical range, of infected plants, and in adopting it widely man has perhaps played his biggest part in dispersing pathogens. The spread of viruses is particularly important because many varieties of plants carry virus without showing symptoms. Fungal resting spores are also probably transported by man over long distances, as are infective insect vectors, but little is known about such spread; it is probably less important, compared with the moving of infected plants, because the pathogen has few chances of surviving and spreading to other susceptible plants.

Dispersal of a pathogen on a worker's body or clothes, or on machinery or draught animals must usually be local, in contrast to man's dispersal of infected tissues over long distance. Few viruses are transmitted in expressed plant sap outside the laboratory because they are rarely concentrated enough to persist and cause infection. Tobacco mosaic virus, however, is so stable that it can infect after 2 years on equipment and potato X virus can persist for up to 6 weeks on clothing, so these can spread from one crop to another, as well as within crops, during cultivations or inspections. Tobacco mosaic virus is spread widely among tomatoes while the plants are being tied up or side shoots are being removed; it is so persistent that it can survive when plants are composed or when tobacco is cured, and may be introduced into an uninfected tomato or tobacco crop on the hands of smokers. Potato spindle tuber virus is spread by the knives used to cut tubers into seed pieces, a prevalent practice in America, and also by tractors passing through healthy potato crops after infected ones. Tulip flower-break and *Cymbidium* mosaic are 2 other viruses spread on cutting knives, but *Narcissus* mosaic is not; neither is the infectious potato X virus, possibly because tuber parenchyma has too low a virus content to be a source of inoculum and also resists infection by inoculation. Fruit growers who graft infected budwood scions, on to rootstocks provide an efficient and for some viruses of top-fruit, the only known means of spread.

Bacteria, too, can be spread on cutting knives: *Corynebacterium sepedonicum* (Spieck. and Kotth.) Skapt. and Burkh, which causes ring rot of potato, spreads rapidly when stocks with a proportion of infected seed tubers are cut and planted immediately, but less readily when the cut surfaces are allowed to dry before planting. The sacks used to store or transport potatoes can also provide sources of infection. Several other bacteria are spread on knives; for example, *C. michiganense* (E.F. Smith) Jensen, causing bacterial canker of tomato, and *Pseudomonas solanacearum* E.F. Smith, causing bacterial wilt of bananas.

Many fungal spores are disseminated on hands, clothing, and machinery, but they are relatively unimportant in the epidemiology of the diseases, and there has been little work on this types of spread. Fungi, and soil-borne pathogens of all kinds, can easily be transported on boots, implements, and the roots of transplanted plants; two well-known examples are *Plasmodiophora brassicae* Woron., causing clubroot of crucifers, and *Synchytrium endobioticum* (Schilb.) Perc., causing potato wart disease—fungi that, although they cannot live as saprophytes in the soil, can exist as resting spores for many years in the absence of host plants. *Verticillium albo-atrum* Reinke and Berth, causing wilt of hops and many other plants, persists for only a short time as a saprophyte in soils and is spread mainly by diseased plant residues carried alone or in soil. Wilt often follows the lines of cultivation down the rows of hop gardens, and is also spread around by the hop pickers.

II. DISPERSAL BY OTHER MAMMALS AND BIRDS

Very little is known about the dispersal of pathogens by wild animals. Dispersal by draught animals attached to cultivation machinery can reasonably be attributed to man. Pathogens that can persist on clothing or machinery might easily be carried on an animal's hair or a bird's feathers, and rabbits and dogs spread potato X virus in potato crops when they brush past infected plants and break the leaf hairs. Birds have not been described as vectors of plant viruses, but there seems no reason why they should not carry some on their feathers or beaks. Game birds, such as partridge, often run within crops before flying and might well be responsible for some of the unaccountable introduction of virus X into healthy potato crops.

Dogs, mice, birds, and indeed, frogs, and anything else that moves in flax crops when the plants are wet spread *Sphaerella linarum* Woll., spores of which ooze from infected flax plants in a gelatinous matrix. Some of these animals may also spread the fungus to healthy crops over a distance, and small mammals and birds may also disseminate other similar fungi in this fashion. Birds carry the chestnut blight fungus, *Endothia parasitica* (Murr.) Anderson and Anderson; this spreads locally by air-borne ascospores discharged from perithecia in wet weather, but outbreaks in new areas and the consequent rapid spread through North America were probably caused by insectivorous birds. Old cankers of chestnut blight are often infected with boring insects, for which birds, especially woodpeckers, search and then become contaminated with pycnospores. Several birds, shot while in the branches of infected trees, were carrying *Endothia* spores; two woodpeckers each had more than half a million spores on its plumage. Leach (1940) pointed out that woodpeckers feed on tree cambium as well as on insects, and so could easily infect healthy trees during feeding.

Little work has been done on the passage of pathogens through the intestines of animals other than insects. Many nonpathogenic fungi are coprophilous, and some are adapted to passage through animals. The resting spores of two plant pathogenic fungi are spread in dung to clean fields, after potatoes infected with *S. endobioticum* (wart disease) or brassica plants infected with *P. brassica* (clubroot) have been fed to stock.

III. DISPERSAL BY SMALL ANIMALS, OTHER THAN INSECTS

The eriophyid mite *Phytoptus ribis* Nal. has long been known to transmit black currant reversion virus, but recently eriophyids were also found to be vectors of several other viruses. Wheat streak mosaic and wheat spot mosaic viruses are transmitted by all active stages of *Aceria tulipae* Keifer that are reared on infected plants. The mites remain infective for several days and through molts; adults are unable to acquire virus, but nymphs become infective after 30 minutes on infected plants. Wheat streak mosaic virus is carried by wind-borne mites, which breed mainly on wheat, and do not survive long off living plants. Several species of grass are susceptible to the virus, but the disease is found on wheat only when sown near other wheat crops or volunteer wheat; plants that emerge after adjacent crops have matured are not infected. Mites also transmit fig mosaic and peach mosaic viruses; they inhabit closely adhering peach leaf-bud scales, which are characteristic of the retarded buds of infected trees in summer, so the virus apparently modifies the tree in favor of the vector.

Mites have seldom been shown to transmit bacteria or fungi; since they usually occur in rotting tissues of various bulbs and of carnation buds, it is assumed that they transmit the pathogens responsible, but this needs confirming. They also often occur in the tunnels formed by bark beetles, and may possibly play a secondary role in spreading Dutch elm disease and blue-stain fungi. As Leach (1940) points out, many of the mites are carried by flies and other winged insects that feed on decaying tissues. These insects themselves might carry pathogens, but the fact that mites are wingless does not preclude them from being vectors. Nevertheless, more critical experiments are needed before their significance as vectors can be assessed.

Leach also quoted a few papers which show that slugs occasionally transmit fungus spores from diseased to healthy plants, but they are probably of little importance as vectors.

Like some insects several nematode species eject the contents of the esophageal gland when feeding and in doing so might transmit pathogens. Indeed, Steiner (1942) concluded that soil nematodes are important as vectors of bacteria and viruses, but no virus, soil-borne or other, has yet been shown to be carried by nematodes. No doubt pathogens enter through nematode feeding punctures, but although various associations between nematodes and pathogen have been described, little critical work has been done on them.

A few bacterial and fungal diseases seem to be more prevalent when nematodes are common than when they are not, and some depend on nematodes for their spread. One bacterium that apparently infects only when carried by nematodes is *Corynebacterium fascians* (Tilf.) Dowson, which, with the eelworms *Aphelenchoides ritzema-bosi* (Schwartz) Steiner and *A. fragariae* Christie, causes the hyperplastic "cauliflower" disease and enations of strawberry. The nematodes are ectoparasitic and carry the bacteria to the meristematic tissues, through which the plant is infected, but which are normally inaccessible to bacteria because they are tightly enclosed within stipules. Similarly, *C. tritici* (Hutchinson) Burkh. does not infect wheat in the absence of *Anguina tritici* (Steinbuch) Filipjev. Galls containing nematodes (cockles) are formed in place of seeds, and are then dispersed by man; bacteria carried within them can remain viable for months. The same eelworms are almost always associated with the fungus *Dilophospora alopecuri* F., which causes leaf spotting and deformed heads of wheat and other cereals. The nematodes emerge from the cockles in moist soil and live as ectoparasites on cereal plants until they invade the ovaries. The fungus infects the nematode galls and its spores become attached by bristle-like appendages on the mucous covering of the eelworms, which can also acquire spores from germinating pycnidia in the soil. The spores appear to infect only when placed in wounds near the growing point.

Nematodes even though not spreading pathogens may nevertheless help them to infect plants. For instance, several root knot nematodes increase the incidence of carnation wilt by providing wounds for the entry of *Pseudomonas caryophylli* Starr and Burkh, into roots.

IV. DISPERSAL BY INSECTS

After the discovery in the mid-19th century that fungi and bacteria cause many diseases, it was sometimes postulated that insects might disperse them, but the first experimental evidence was not obtained until 1891, when Waite showed that bees and wasps carry *Erwinia amylovora* (Burrill) Winslow et. al., the fire blight bacterium of pears, from diseased to healthy blossoms while searching for nectar. By 1920 several virus, bacterial, and fungal diseases were known to be caused by pathogens transmitted by insects, but relatively little work was done on the subject because plant pathologists and entomologists rarely worked together and seldom trespassed into one another's subjects. Later, viruses attracted workers from both disciplines, and consequently more is now known about their transmission by insects than that of other pathogens. Leach and his co-workers in the United States have pioneered work on the spread of fungi and bacteria by insects, and Leach's book (1940) has stimulated interest in this, partly because some of the diseases, such as Dutch elm disease and chestnut blight, are of considerable economic importance.

Insects are often associated with fungal and bacterial diseases, but it is sometimes difficult to prove that they are vectors. They are, however, well equipped to act as such, because most of them depend on plants for food, they are generally active, and their body bristles enable them to carry many spores or bacteria externally. In addition, many plant pathogens can survive, and some even multiply, inside insects, and insects that regurgitate during feeding seem particularly well adapted to act as vectors.

A. Method of Transmission

1. *Viruses*. Not all insects that feed first on a diseased plant and then on a healthy one transmit viruses. Virus must be introduced into a living cell before infection occurs, and so the Homoptera, with piercing and sucking mouth parts, are particularly effective vectors. Most of them feed without causing major damage to plant cells, unlike the Heteroptera, which often have toxic saliva. Damage around the feeding puncture may be one reason why so few Heteroptera transmit viruses.

Some viruses are rarely or never insect-transmitted; some are transmitted by one or a few species of insect; and others by many, but no claim that a virus is transmitted by more than one major group of insects has yet been confirmed, except where the virus is carried as a contaminant on the body, as is tobacco mosaic virus by leaf-miner flies. This virus can also be transmitted occasionally by grasshoppers. Some of the differences in behavior depend on the virus, for the same insect species may show very different efficiency in transmitting different viruses.

There is at present no satisfactory system of virus classification, but insect-transmitted viruses can be divided into three imprecise groups according to their behavior in the vector: (1) viruses that are acquired in brief feeding periods (often less than a minute) on infected plants; the vectors usually cease to be infective within an hour, and the efficiency of transmission is increased if they are prevented from feeding for a few hours before feeding on the infected plant. This group comprises the "nonpersistent" viruses of Watson and Roberts (1939). (2) "Persistent" viruses are acquired only after longer feeding periods on infected plants, and once infective, the insect vector often remains so for many days or weeks. (3) some viruses, such as beet yellows, have vectors that remain infective for several hours, and have been called "semipersistent."

Nonpersistent viruses are transmitted by aphids, which also transmit some persistent ones. All those known to be transmitted by leaf hoppers, white flies, bugs, thrips, and most biting insects are persistent. They usually have a noninfective or "latent" period, that is, the insect is unable to transmit virus for some hours or days after feeding on an infected plant, whereas nonpersistent viruses are transmitted immediately after the acquisition feed. Some viruses that are transmitted only after a noninfective period in the vector may need to multiply within the insect to reach a transmissible amount, but with others the period may simply be the time needed for virus to pass through the gut wall into the blood and thence to the salivary glands. These noninfective periods differ greatly in length with different viruses and insects.

Several leaf-hopper-transmitted viruses multiple within their vectors and some are transmitted through the eggs, which provide the viruses with an alternative means of survival than in infected plants. But not all persistent viruses multiply and remain infective within the insect, and beet leaf hoppers, *Circulifer tenellus* lose the capacity to transmit beet curly top virus as they age.

Many viruses are transmitted by one or very few leaf hopper species, but the causes of such specificity are largely unknown. In some it may be connected with the insect's feeding habit and the distribution of the virus within the plant; for instance, beet curly top virus seems to be restricted to the phloem, and lucerne dwarf virus to the xylem. Vectors are then restricted to phloem or xylem feeders, and a reason why insects often transmit virus only to some plants in a series may be that the hopper does not always reach the tissues in which the virus can develop. Some jasside are known to find the phloem more readily than others, possibly because they follow a pH gradient from the epidermis to the phloem.

All instars of most insect vectors are able to acquire and transmit virus, but sometimes the noninfective period of persistent viruses is longer than the development period of larvae or nymphs, and then only adults can be vectors. Tomato spotted wilt virus, however, cannot be acquired by adult thrips, but these can transmit virus when they acquire it as larvae. Some instars may be more efficient vectors than others; for instance, adults of the mealy but *Pseudococcus citri* (Risso) transmit cacao swollen shoot virus more readily than young nymphs.

Most persistent viruses are not readily sap-transmitted, possibly because they are in too low concentration in plant extracts or because they need to be put into specific tissues, but those transmitted by biting insects are unusual in persisting in their insect vectors for considerable periods and also being easily sap-transmissible. Flea-beetle vectors of turnip yellow mosaic virus and the cucumber beetle vectors of squash mosaic virus regurgitate infective juice from the foregut during feeding; they not only infect plants fed on immediately after an acquisition feed, but continue to be infective for several days.

The noninfective period in aphid vectors of persistent viruses differs with different test plants and environments. When potato leaf roll virus is acquired from potato by *Myzus persicae*, the aphid is not infective for about 2 days, but this period is decreased to several hours when the source is *Datura stramonium* L. and sometimes to less than an hour with *Physalis floridana* Rybd. Although it might be expected that some persistent viruses multiply in aphids, as in leaf hoppers, the evidence for potato leaf roll and other viruses is inconclusive.

Because aphids transmit nonpersistent viruses much more readily when starved before the acquisition feed, those that have flown a long way are in a favorable state to transmit if they feed on an infected plant when they land in a crop. Some of these viruses can be acquired or transmitted extremely quickly, often by aphids that probe the epidermis to test the suitability of a plant as a source of food; thus potato Y virus can be acquired in 5 seconds, but probes lasting up to 1 minute are more likely to make an aphid infective. When the aphids' stylets remain inserted for 20 minutes or longer, most aphids are not infective, and many cease to be infective within 15 minutes of leaving the infected plant, even when fasting. Virus is carried near the tips of the stylets and aphids rarely become infective after the stylets penetrate beyond the first layer of cells. The stylet tips are well-adapted for carrying virus-containing sap, and van Hoof (1958) suggested that virus might be acquired more readily from epidermis than from parenchyma because aphids pierce the epidermis through the cell wall, but the parenchyma through the intercellular spaces. Both he and Bradley found that aphids could acquire virus from the parenchyma, so the suggestion of Bawden et. al. (1954) that nonpersistent virus occurs predominantly in epidermal cells may not be the only reason for the effectiveness of short acquisition feeds.

Although many nonpersistent viruses are transmitted by several species of aphid, there is still considerable vector specificity. Thus *Myzus ornatus* Laing, *Myzus ascalonicus* Donc., and *Aulacorthum solani*, (Kltb.) transmit dandelion yellow mosaic virus and not lettuce mosaic, whereas *M. persicae* transmits lettuce mosaic virus but not dandelion mosaic. Even when several species can transmit, some do so more readily than others. The cause of such differences has not been determined although the presence of different virus inhibitors in the aphid's saliva or different abilities to adsorb viruses onto the stylets have been postulated.

Aphid populations are usually heterogeneous and individual vectors often differ in their efficiency; some strains of *M. persicae* are very poor, some very good vectors of a particular virus, but it is not known whether an efficient strain for one virus would also be efficient for others. Not all individuals of the leaf hopper *Cicadulina mbila* Naude transmit maize streak virus. Those that cannot (the inactive) pick up the virus while feeding on an infected plant, but virus does not get into their blood. Inactive insects become able to transmit virus after their intestine is pierced by a needle to allow virus to enter the blood. Attributes of the virus particles that are changeable seem to determine transmissibility by a given species of aphid. The spinach strain of cucumber mosaic virus was transmitted readily by *M. persicae* during 1945–1954, but not at all in 1957, although it was still transmitted as regularly by 2 other species of aphids as in earlier years. Other strains of this virus were as readily transmitted by *M. persicae* in 1955. Other viruses have ceased to be insect-transmitted after being propagated by mechanical transmission for several years, and an instance of gaining transmissibility is provided by potato C virus, which was not aphid-transmitted when derived from potatoes in 1945 and 1955, although apparently closely related to the aphid-transmitted Y virus. After propagation by mechanical inoculation in *Nicotiana glutinosa* L. and *N. tabacum* L. for 10 years it could be transmitted by *M. persicae*. However, when passed through potato and back to tobacco, it ceased to be aphid-transmitted. This suggests that the attributes of a virus that allow it to be insect-transmitted can be affected by the host plant in which it is multiplying.

2. *Bacteria*. Unlike viruses, very few of which are carried externally as contaminants by insects, most insect-transmitted bacteria are carried incidentally. A few insects, which probably have been long associated with the bacteria concerned, appear to have a symbiotic relationship with them, and still fewer seem specially modified morphologically to carry bacteria and ensure their survival.

Like all bacteria carried incidentally *Erwinia amylovora*, which causes fireblight of apples and pears, is not dependent on insects, as both primary and secondary infections occur by rain splash. Nevertheless, insects play a prominent part in dissemination, and primary infection occurs when flies, ants, and others carry bacteria from bark canker exudate to the blossoms. Bees and wasps carry bacteria from flower to flower on their mouth parts, but infection occurs only when the sugar concentration of the nectar is low. Aphids, bugs, and bark beetles carry the bacteria externally and infect shoots when feeding, but bacteria may live for several days in the intestines of flies, and eggs may be contaminated as they are laid; bacteria can persist also in puparia and contaminate emerging adults.

Larvae of cabbage root fly (*Erioishia brassica* Bouché) carry *Erwinia carotovora* Holland, both externally and internally; bacteria are acquired from fly eggs which are contaminated as they are laid or from decaying plant material or the soil, and they can overwinter in the pupae. They cause stump and heart rot of cabbages and other brassicas. Similar pathogens cause soft rot of onions, carried by the onion fly (*Delia antiqua* Meig.) and celery rot, transmitted by two leaf miners.

Erwinia tracheiphila (E.F. Smith) Holland, which causes bacterial wilt of cucurbits, is apparently entirely dependent on the beetles *Diobrotica vittata* Fabr. and *D. duodecimpunctata* Oliv. for overwintering and transmission (Rand and Enlows, 1920). The beetles are particularly effective vectors because during feeding they wound the vascular boundless to which the bacteria are largely confined.

Lepidoptera have seldom been described as vectors of plant pathogens, but the moth *Cactobrosis fernaldialis* (Hulst) is the main vector of the bacterium *Erwinia carnegieana* Lightle *et. al*., which destroys giant, cacti in Arizona. Bacteria can be isolated from adult moths, and egg surfaces and larvae carry them externally and internally. Larvae move away from severely diseased tissues and carry the pathogen through the cork, which would otherwise seal off the necrotic area. When potato tubers are cut into "seed pieces", the formation of cork prevents entry of

Erwinia atroseptica (van Hall) Jennison, causing "blackleg", or seals off necrotic lesions, but insects that feed on the potato below ground aid its entry, and by burrowing in the tubers, prevent the formation of wound cork. Some of these insects have a symbiotic relationship with the bacteria: species of *Delia* develop normally on infected tubers, but not on sterile ones. The bacteria occur in many cultivated soils, but are unable to penetrate undamaged roots or tubers. Bacterial rot of apples, caused by *Pseudomonas melophthora* Allen and Riker, follows feeding by the apple maggot (*Rhagoletis pomonella* Walsh). Bacteria are carried externally and internally by the adults; apples are inoculated when the eggs are laid, and the bacteria are carried in by the burrowing larvae, which prefer rotting tissues.

A close symbiotic relationship has been reported between the olive fly (*Dacus oleae* Rossi) and several bacteria, including *Pseudomonas savastonoi* (E.F. Smith) Stevens, which causes the olive knot disease. Anatomical modification of the adult insects ensures the perpetuation of bacteria through successive generations: the anal tract of the female contains bacteria in several sacs, opposite which is a slit in the membrane separating the tract from the vagina. When an egg passes along the vagina, the slit opens and the egg presses against the openings of the sacs; some bacteria enter the micropyle and later the developing embryo. They are inserted into oviposition wounds, which, according to Petri, are the source of most olive knot in Italy. The insect is not present in California, and there the bacteria spread more slowly.

These examples illustrate two important facts. First, few bacteria cause systemic infection. Insect vectors often not only make the initial infection, as with viruses, but also often spread the bacteria from place to place over plants. Second, many bacteria are carried internally by insects, a fact which may be of evolutionary significance in helping to tide them over unfavorable periods.

3. *Fungi*. Some fungi differ from viruses and bacteria because they can actively penetrate plant epidermis. Others depend on natural openings such as stomata, while still others, such as bacteria, enter through wounds. Relatively few fungi are spread by insects, but some are so well-adapted to insect transmission that they may have been associated for a long time.

Some insects, especially pollinators which feed without wounding the plant, act merely as vectors of fungi, but the pathogen is often well-adapted for this form of spread. For instance, *Botrytis anthophila* Bond., spread by bees, sporulates only on the anthers of red clover, and *Ustilago violacea* (Pers.) Roussel, spread by nocturnal moths, replaces the pollen of campions, pinks, and other Caryophyllaceae with sticky smut spores; the petals are unaffected and the flowers remain attractive to insects. Mycelium penetrates the ovary but does not destroy the developing seed, which later produces a systemically infected plant whose flowers all produce spores. A small hymenopteran, *Blastophaga psenes* (L.), ensures an unusual method of pollination called caprification and causes an internal rot of figs by carrying spores of *Fusarium moniliforme* var. *fici* Caldis from infected to healthy fruits. The adult male insects emerge from their galls and penetrate to the female insects in their galls in the male inedible figs. The fertile females emerge through the eye of the fig and so collect pollen from the staminate flowers. Male figs, some of which may be infected with the fungus, are hung on female trees when the insects are due to emerge; these enter the receptacles of the female edible figs to oviposit, pollinating and inoculating the fruit with fungus while doing so. They cannot oviposit in the female fig, however, because the styles are much longer than those of the male flowers. Without pollination the fruit of some varieties drops before it is mature.

Some insects carry fungi on their mouth parts and introduce them into plants when feeding. *Nematospora gossypii* (Ashby and Now.), causing stigmatomycosis of cotton, probably depends entirely on cotton stainers and related bugs (*Dysdercus* spp.) which introduce into the bolls needle-like ascospores that are thin enough to allow passage through the stylet canals to and from the stylet pouches of the insect. Plant bugs are noted for the injury they cause when feeding, and Leach (1940) suggests that the necroses may not all be caused by toxic saliva, but may often be caused by associated fungi of the *Nematospora* type.

Other fungi are carried into the plant by boring insects; the bark beetles that carry *Ceratostomella ulmi* Buism., causing Dutch elm disease, are mainly in the genus *Scolytus*, and fungus sporulates freely in their egg galleries. Adults are covered by, and ingest spore-containing slime with which they inoculate the twigs of healthy trees, upon which they prefer to feed before breeding beneath the bark of trees that have been weakened by the fungus during previous years. Several brown- and blue-stain diseases of conifers are caused by species of *Ceratostomella* and related fungi, which stain but do not rot the wood and are similarly transmitted by bark beetles. These beetles (*Ips* and *Dendroctonus* spp.) breed only in trees that are dead or weakened by other agents; they do not weaken trees by inoculating them with fungus, as do *Scolytus*. When they attack living trees, however, the fungus blocks the transpiration stream and the trees soon die. Although fungi are ingested by the larvae, Leach (1940) considers them unimportant as food, but important because they weaken the tree, decrease its water content, and make it more suitable for beetles. Three species of scolytids infest different parts of white fir (*Abies concolor* Lindl. and Gord.) in California; the fungus *Spicaria anomola* (Corda) Harz. is associated with those at the top of the tree and in the branches, and causes a light brown stain. *Trichosporicum symbioticum* Wright causes a darker stain and is associated with a third species at the bottom of the tree, although both fungi will grow anywhere on the tree when inoculated. Both kill the cambium as they advance, as do many scolytid-carried fungi, preventing an inflow of resin into the brood galleries of the beetles.

The importance of these beetle vectors to the lumber industry was demonstrated by Verrall (1941). Since chemical fungicide treatment of lumber became common, dispersal of air-borne spores has become unimportant, but the ambrosia beetles that attack hardwoods and the bark beetles that attack softwoods are undeterred by the protective surface of chemicals and inoculate the timber below it. Boring insects are not confined to wood. Larvae of the corn borer, *Pyrausta nubilalis* Hübn., distribute several pathogenic fungi and bacteria within maize plants. They weaken adjoining tissues and make them more susceptible to fungal attack, and their frass provides medium for fungal development.

A fungus that enters through feeding wound is *Chilonectria cucurbitula* (Curr.) Sacc., which causes burn blight of pines. Numerous spores exude whenever mature perithecia are moistened, and they are common on the foliage of attacked trees. Adult spittle insects (*Aphrophora saratogensis* Fitch) are contaminated as they crawl about on small twigs near the tops of trees during the autumn, and the fungus enters through their feeding punctures. The next year the fungus girdles the twigs, then moves down the cortex, killing the tree during the next 3 years. Wasps and other insects spread *Sclerotinia fructigena* Aderh. and Ruhl. when they feed, for spores usually enter only through wounds. The fungus spreads rapidly through fruit, causing brown rots of apples, pears, plums, and peaches, and sporulates on the surface, so the insects readily become contaminated.

Feeding is not the only way in which insects wound plants, and some fungi are introduced when eggs are deposited in plant tissues. *Urocerus gigas* (L.) and *Sirex cyaneus* F., the wood wasps, inoculate *Stereum sanguinolentum* (Fr.) Fr. in conifers and cause heart rot. The insects are well adapted for this, and carry the fungus in sacs at the anterior end of the ovipositor, so the egg is contaminated as it is laid. Larvae eat hyphae and carry the fungus in their hypopleural organs. A less intimate relationship exists between crickets of the genus *Oecanthus* and *Leptosphaeria coniothyrium* (Fuckel.) Sacc., causing tree-cricket canker of apple and cane blight of raspberry. The female eats as small hole in the bark, deposits her egg, and then closes the hole, either with a fecal pellet or a chewed piece of bark, usually diseased. Spores and hyphae of this and many other fungi are carried internally and externally by the insects and contaminate the wound.

In an entirely different category of vectors are the fungus feeders, mainly Diptera and Coleoptera, which feed on the sweet sticky secretions that contain the spores; they are contaminated by these, and later leave them on healthy flowers or plants. Examples are the *Sphacelia* stage of the ergot fungus, *Claviceps purpurea* (Fr.) Tul., and the pycniospores of thistle rust. Wind-borne ascospores of *C. purpurea* infect rye and other Gramineae in spring; the fungus invades the ovary and develops conidia (*Sphacelia*) in a secretion which is seemingly attractive to insects. Several insects visit rye to feed on the pollen and secretion, and are contaminated externally, or they ingest the spores, which may be excreted or regurgitated later on a healthy spikelet. Craigie (1927) showed the importance of flies in producing "hybrids" of rust fungi by carrying pycniospores from one pycnidium to another.

Fungus feeders that have caused considerable economic loss during recent years are the nitidulid and scolytid beetles that spread the oak wilt fungus, *Endoconidiophora fagacearum* Bretz. This fungus forms mycelial mats under the bark, which splits above them; beetles are attracted, presumably by the odor, and both they and their larvae feed on the fungus. The fungus is heterothallic, and the beetles serve as spermatization agents by transmitting endoconidia from mats of one type to those of the other, thus stimulating the production of perithecia and ascospores. The two thallus types are rarely intermingled and depend on insects for cross fertilization. The insects also inoculate fresh wounds with spores are they walk over them.

B. Geographical Distribution

The spread of plants around the world has widely distributed many pathogens and vectors; these are not always carried simultaneously to a new area, and a pathogen introduced alone will not spread unless a "local" insect can act as a vector. Thus tristeza disease of citrus trees became prevalent in South America soon after infected trees were imported from South Africa because *Aphis citricidus* Kirk, an efficient vector, was already prevalent. As this aphid does not occur in California, tristeza spreads less rapidly there than in South America.

Viruses may be introduced into a new area unwittingly in insectfree and apparently healthy plants, and may then be transmitted by new vectors to plants which react with a severe disease. A newly introduced insect may also cause trouble by transmitting an indigenous virus from plants which are little affected to others which react severely.

The discovery of Dutch elm disease in the Netherlands in 1919 and the rapidity with which it assumed the status of a major disease suggests that the fungus had not long been associated with bark beetles, which had been known for centuries to infest weakened trees. It was probably not widespread in Europe earlier because the disease is obvious and also, when *Scolytus multistriatus* Marsh. was taken to the United States before 1909, the fungus was not introduced. A later introduction in the 1920's included the pathogen, and this beetle is now the most important vector in the United States. *Ceratostomella ulmi* might have replaced less efficient fungi with which the beetles were originally associated although, as there are nonpathogenic strains, a pathogenic one may have arise by mutation. Similarly, saprophytic fungi are always associated with nitidulid beetles in wounds on healthy oak trees where they serve as food for both adults and larvae. They have not been found without the beetles, with which they are apparently symbiotic. Jewell (1956) suggested that *Endocondiophora fagacearum* had recently become a symbiont, for oak wilt does not occur in many areas where both oaks and beetles abound.

As in organism, mutation in viruses, coupled with geographical isolation, appears to lead to the development of distinct strains. Thus Brazil, Argentina, and North America have their characteristic strains of beet curly top virus, each with a different species of leaf hopper as vector. A strain of curly top similar to the North American strain occurs in Turkey, and Bennett and Tanrisever (1957) postulated that the virus originated in Europe and was carried with beet to several parts of the New World, where it acquired different vectors.

Some polyphagous insects are restricted to a few plants during part of the years thus *Myzus persicae* overwinters only on *Prunus* spp. in many areas and on horticultural crops in others. Although the aphids cannot acquire potato viruses from such plants and some fly long distances, the potato virus diseases are often most prevalent in crops near peach orchards or gardens because aphid vectors are most numerous there.

Changes in cropping can affect the number of vectors in a specific area: millions of *Prunus serotina* Ehrh. were planted in the north Netherlands as forest shade trees during the 1940's and proved to be excellent winter hosts for *M. persicae* in an area where peach is scarce. This species also increased greatly in the Imperial Valley of California when the amount of sugar beet was increased. This in turn increased the incidence of cantaloupe mosaic in melons even though the aphids do not colonize the melons. Leaf hopper vectors of beet curly top virus multiplied greatly during the depression of the 1920's on the weeds of abandoned farms in the western United States. The natural sagebrush, and well-cultivated grass, are not favorable covers for hoppers, but overgrazing during this period turned many ranges into semidesert in which the hoppers flourished.

Predators and parasites play a major part in altering an insect population from one year to another and also from one area to another. It is often difficult to assess their influence because so many insects are involved. Few studies have been made in the detail attempted by Hille Ris Lamber (1955), who found that the aphid-predator-parasite-hyperparasite population in a potato crop included over 50 species. After aphids have become numerous on crops such as potato, they often suddenly disappear. Hanesen (1950) and Hille Ris Lambers noted that this decrease occurs earlier when aphids are more numerous than usual, and they attributed this to insect enemies. While agreeing that parasites and predators always help to determine the ultimate size of the population, other workers consider the major cause of the decline to be the departure of winged aphids. As potato plants mature they become less suitable sources of food for aphids, most of which become winged and fly away. This often occurs when parasites and predators are most numerous, so remaining aphids are quickly eliminated. Occasionally, however, the enemies are very numerous early, when aphids are colonizing the potatoes, and then a large summer aphid population may fail to develop. In many parts of Europe there is a tendency toward a biennial rhythm, a year with many aphids being followed by one with few, because predators and parasites multiply greatly in seasons when aphids are numerous during the summer, and may then overwinter and help prevent the aphid infestation from developing the following spring. Only when their enemies have decreased in number from lack of food can the aphids again multiply unchecked.

The influence of parasite distribution over a wider area was noted by Stubbs (1956), who contrasted the rapid spread of carrot motley dwarf virus in Australia, where the vector, *Cavariella aegopodii* (Scopp.) is very numerous, with the slow spread in California, where the aphids are few because they are severely parasitized. He suggested that the vector is more in equilibrium with its environment in California than in Australia, where it may have been introduced rather recently.

C. Seasons, Climate and Weather

Geographical differences in vector populations are usually determined by the differences in climate if the requisite plant hosts and the vector have been widely distributed. Climate affects the seasonal cycles of insects, but their numbers and activity also differ from one year to another because of differences in weather. The optimum temperature for aphid reproduction is about 26°C. Consequently they are more numerous in continental than in maritime climates, and in warm, dry summers than in cool, wet ones. Cultural practices vary from one country to another, so it is difficult to relate the incidences of virus diseases in different parts of Europe, for example, to differences in aphid numbers, although differences in the same area can be related to the vectors. Weather plays a large part in regulating outbreaks of aphids; rain and cold restrict larval development and, consequently, the number of adults. Fewer winged aphids develop in wet weather, and because these seldom fly when it is cool fewer new colonies are founded than when the weather is warm and dry.

In temperate climates emphasis tends to be placed on the regulating effect of cold weather, but aphids are soon killed when temperatures rise a few degrees above the optimum. Van der Plank (1944) found that *M. persicae* forsakes potatoes in Africa, and viruses cease to spread when the mean mainly maximum temperature reaches 32°C. The adverse effect of a hot climate on aphids is also used to produce virus-free lettuce seed.

In an arid climate rain may have an effect opposite to that in a cool one: outbreaks of Pierce's disease of grapevines in California are most severe during wet periods because host plants of the leaf hopper vectors grow best then.

Humidity also determines the loss in celery from the bacterium *Erwinia carotovora*: the larval leaf miners that inoculate plants remain in the outer leaves during wet weather, and soft rot causes relatively little loss, but when it is hot and dry, the flies deposit eggs near the heart of the plant. Larvae search for a moist place when they hatch, and enter the young leaves causing heart rot.

Some fungi produce spores in a gelatinous matrix only during very humid or wet weather; spores of *Sphaerella linarum* and others that are carried as contaminants on animals' bodies can only be spread, therefore, when the foliage is wet, a condition when infection, too, is more likely to occur than when it is dry, for most spores need a high humidity to germinate and infect.

The seasonal cycle of insects depends on the climate, which often differs considerably in regions not widely separated; in Great Britain, for example, most aphids occur on potatoes during July in the southern half of the country, during August in northern England, and during September in parts of Scotland. The production of seed tubers free from leaf roll and Y viruses in Scotland was unwittingly based on this. The principal vector of these potato viruses, and of many others of economic importance in Europe and elsewhere, is *Myzus persicae*; most potential vectors are usually present in England during July and early August, but virus is not necessarily spread mostly at this time, except from one crop to another. A rapid increase in disease incidence often depends on the activity of vectors within crops when plants are very susceptible to infection, such as when relatively few aphids colonize potato crops in the spring.

The time of maximum population of some aphids depends on their plant hosts as well as on climate. Thus the strawberry aphid *Pentatrichopus fragaefolii* (Cock.) is most numerous in late summer on first-year plants but in late May or June on older ones. Winged forms are numerous only when the population is maximal, and most virus spread coincides with their activity. On the other hand the apterae (wingless forms) are also numerous at these times. On that account Posnette and Cropley (1954) could not determine which were principally responsible for the spread of virus to adjacent plants. The growth of plants in a crop may so modify the microclimate that the vectors can no longer breed on them. *Circulifer tenellus* infests beet and spreads curly top virus within the crop only when the plants are young because the environment is too humid for the leaf hoppers when the plants cover the soil.

Many insects are more active and may carry virus farther when it is warm than when it is cool. Bald (1937) recorded a close positive correlation between temperature, thrips activity, and the number of plants showing spotted wilt 12 days later. In Florida aphids carry pepper veinbanding mosaic virus to peppers only within 150 ft. of the source when the temperature averages 62°C., but much farther at higher temperatures. In addition to its effect on movement high temperature may affect the insect's infectivity and the plant's susceptibility: *M. persicae* is a more efficient vector of potato leaf roll virus when reared on infected plants at 27°C. than at 22°C., and the resistance of the potato is lower at 27°C.

D. Availability of the Pathogen and Susceptibility of the Host

Virus in an infected plant may be more readily available to a vector at one time than at another. Young plants are usually the best sources of virus became the concentration of many viruses decreases as the plants cease to grow rapidly: thus *M. persicae* transmits potato leaf roll virus to few test plants when the source is old, glasshouse-grown infected plants, but readily from very young ones.

The distribution of virus within the plant can determine when insects acquire virus and on which parts of the plant they need to feed to do so. Many viruses can be acquired by insects some days before a newly infected plant shows symptoms, for example cauliflower mosaic. Kato (1957) found, however, that Y virus can be recovered by aphids from potato only after symptoms show. Aphids can acquire potato leaf roll virus from the lower leaves of full-grown potato plants much more readily than from the rest of the plant, suggesting that virus concentration is not always greatest in the youngest, fast-growing leaves. That unequal distribution is sometimes an attribute of the virus, not of the host, was shown when the spread of 2 viruses in cauliflower crops was studied. Both cauliflower mosaic virus and cabbage black ring spot virus are transmitted mainly by *M. persicae* and *B. brassicae*, and both viruses spread readily when the infected source plants are young; but when they are old, cabbage black ring spot virus spreads less readily than mosaic virus, which occurs in high concentration in all the new leaves produced by infected plants, whereas ring spot virus accumulated mainly in the older, lower leaves, and even there is localized in the parts that show symptoms. Only in recently infected plant does ring sport virus occur in young leaves in sufficient concentration to be acquired by aphids. As most aphids alight on the upper parts of plants, they are more likely to acquire cauliflower mosaic virus than ring spot virus. Different plant species also differ in their effectiveness as sources of the same virus. Thus, although pepper is a better host plant than chard for *M. persicae* and is more susceptible to southern cucumber mosaic virus, aphids acquire virus more readily from chard than from pepper. Not only does the host affect insect-transmissibility, but the virulence of a virus may be changed by passage through different hosts; Wallace and Murphy (1938) reported that beet curly top virus is less virulent in sugar beet after passage through its wild hosts.

A host plant need not be susceptible to infestation by the vector to be susceptible to infection by a virus; nevertheless, colonizing insects are usually more prevalent in crops than transient visitors, so it is not unreasonable to consider the colonizers first when seeking vectors. It is true that insects are often more active within a crop when they vainly seek a suitable host plant, but colonizers have to be active at some period if they are to find new hosts, and their potentiality as vectors often depends on the readiness with which they move again after landing on a host plant. *M. persicae* was identified early as the principal vector of potato leaf roll and Y viruses; experiments showed that *M. solanifolii* and *Aphis nasturtii* are also efficient vectors of Y virus, and *A. nasturtii* of leaf roll, but field studies show that almost all the spread of the persistent leaf roll virus can be attributed to *M. persicae*, perhaps because of the relative inactivity of *A. nasturtii* after settling on the plants.

These conclusions are confined by the distribution of the virus diseases in Scandinavia. *M. persicae* is confined to the southern coastal areas of Norway and Sweden and so is leaf roll. *A. nasturtii* and *M. solanifolii*, however, occur further north, where Y virus spreads (Lihnell 1948). In brassica crops, also, in Britain, the colonizing *M. persicae* and *B. brassicae* seem to be the only important vectors of cauliflower mosaic virus although at least 20 other noncolonizing species can transmit the virus.

Many workers doubted if an insect species could be the important vector of a virus when it is rarely numerous on the crop because they failed to appreciate the importance of the winged forms. Some viruses, usually nonpersistent ones, are spread mainly by noncolonizing insects which bring virus with them from the plants they have just leaf, or acquire it from infected plants within the crop as they move from plant to plant seeking suitable hosts, as *M. persicae* spreads cantaloupe mosaic among melon crops in California. It is often difficult to find which of the insects that infest or transiently feed on plants spread a virus; or, if more than one species can transmit, it is difficult to assess their relative importance. The principal vector may be the least prevalent insect pest, as is the case in the citrus groves of California where the main vector of tristeza virus, *Aphis gossypii* (Glover), forms only about 3% of the aphids visiting trees.

Even among vector species it cannot be assumed that all the insects that feed or bread on a diseased plant will be infective. Obviously, the more plants that are infected in a crop, the greater will be the proportion of potential vectors that become infective although almost nothing is known about the proportions or numbers of infective aphids in crops. The proportion differs with different viruses and vectors, depending on the insects take to become infective and the time they remain so. Between 10 and 24% of winged *M. persicae* and *B. brassicae* bred on cauliflowers infected with cauliflower mosaic virus are infective when they leave the plants; a similar proportion of *M. persicae*, but fewer than 5% of *B. brassicae*, are infective when bred on plants infected with cabbage black ring spot virus.

Several factors affect the susceptibility of host plants to both insects and pathogens; the most important are variety, age, growth conditions, and population density. Different varieties of crop plants differ not only in the ease with which they become infected, but also in the extent to which the virus multiples in them and so in the readiness with which insects become infective when feeding on them. Varieties may react differently to different viruses; for example, resistance of potatoes to virus Y is not correlated with resistance to leaf roll virus although transmitted by the same aphids. Apparent varietal differences in susceptibility may sometimes be caused by differential feeding by the vectors; thus varieties of lettuce, which experimentally are equally susceptible to yellows virus, contract the disease to different extents in the field.

Susceptibility to infection often decreases with increasing age of plants, and so, other things being equal, incidence of a disease may depend on the age of the crop when infective vectors are active. Beet is most susceptible and intolerant to the leaf-hopper-transmitted curly top virus when in the cotyledon stage; and Hansen (1950) and Steudel (1952) found that numbers of aphids per beet plant and the incidence of yellows increased with successively later sowings. Cereal yellow dwarf virus severely affects only plants infected young, so normally it is of economic importance in California in barley, but not in the earlier-sown oats and wheat. Few plants show such extreme resistance as cassava, however, for although *Bemisia* spp. feed on mature leaves, they infect only the immature ones with mosaic virus.

Some fungi infect only during limited periods, for example, the fungus causing Dutch elm disease spreads readily only during late spring and early summer. Similarly, nitidulid beetles infect healthy trees with the oak wilt fungus only during May and early June, partly because the beetles are attracted to wounds at this time to lay eggs and partly because trees are susceptible only in the spring.

Most workers, who have studied the influence of plant nutrition on the incidence of virus diseases, have found that the best fed plants are the most likely to become infected. Dung and several inorganic fertilizers increase the incidence of both leaf roll and rugose mosaic in potato crops, and aphids also multiply faster on plants treated with dung, sulfate of ammonia, and superphosphate, but less on those treated with muriate of potash. Response to fertilizers varies with the species of aphid, *A. nasturtii* showing little response to the treatments.

Some plants may be more acceptable to the vectors and more susceptible to virus at one temperature than at another. Narcissi are rarely colonized by aphids, and their viruses usually spread slowly, but when retarded bulbs are grown to flower in late summer instead of in spring they are colonized by *Aphis fabae*, and viruses spread rapidly.

Storey (1935) was one of the first plant pathologists to realize that density of the plant population can affect disease incidence: close planting of groundnuts and delayed weeding are practiced by the peasants in East Africa, and greatly decrease the incidence of rosette. Van der Plank and Anderssen (1944) obtained some control of spotted wilt virus, which is brought into tobacco fields soon after transplanting, by increasing the density of plants to 2 or 3 per "hill" and by removing the surplus plants after most of the virus had been brought in. Most insects that bring virus into a crop land at random, so a greater proportion of plants is visited when they are widely spaced than when they are crowded together; consequently, beet yellows, beet mosaic, and cauliflower mosaic incidences are lessened by decreasing the distance between rows or between plants in the row.

Plant size can also affect the spread of viruses because big plants are more likely to be visited by vectors than are small ones, and once infected, the larger plants form bigger reservoirs of virus. In cauliflower seedbeds 30% of the large seedings were infected with cauliflower mosaic virus, in contrast to 15% of medium size seedlings, and 5% of small ones.

Finally, some vectors multiply more rapidly on infected plants than on healthy ones. Several species of leaf hoppers that complete their nymphal stages on celery or aster infected with aster yellows virus die when transferred to healthy plants, but live on diseased ones; to this extent the insects "use" the virus to create a satisfactory source of food for themselves.

E. Introduction of Pathogens into Crops

Persistent viruses must sometimes be carried by insects over hundreds of miles, but it can rarely be proved that this happens and that no local virus sources exist. Occasionally, circumstantial evidence of spread over moderate distances is obtained; for instance, in 1951 *M. persicae* were numerous on leaf roll infected potatoes in the southwest Netherlands, and following southwest winds during the summer dispersal many aphids were trapped about 60 miles to the northeast, where both aphids and virus disease had been scarce; the subsequent outbreak of leaf roll suggested that the aphids had taken virus with them. Because of the difficulty of obtaining evidence, most of the records refer to virus brought into crops from nearby sources. *Macrosteles fascifrons* (Stål) move into lettuce crops from the borders of fields, taking yellows virus acquired from weeds with them; few moved more than 200 ft. during 4 weeks. The rate of vector dispersion, as measured by the incidence of yellows at different distances from the source, differs from one plot to another, probably depending on plant susceptibility, disturbing cultivations, and the weather.

Nonpersistent viruses will rarely be carried far. Observations on the spread of pepper veinbanding mosaic virus from infected *Solanum gracile* showed a steep gradient in incidence of infective peppers, falling from 90% plants infected at 6 ft. to 10% at 50 ft. Nevertheless, a few plants become infected at distances up to 1000 ft. In similar experiments with celery Wellman (1937) found that southern celery mosaic virus was carried by aphids from weeds to over 85% of plants up to 30 ft. away but to only 4% at 120 ft. Distances differed from year to year, but no plant was infected during 3 years in plots 240 ft. away from the source. Storey and Godwin (1953) found that most plants infected with cauliflower mosaic virus occur in the first 50 rows adjacent to diseased crops. Such gradients of disease, from a high incidence in outer rows to a low one within a crop, often serve to show that a pathogen is spreading into a crop from a nearby source.

Taylor and Johnson (1954) studied the deposition of winged *Aphisfabae* and their subsequent multiplication on bean crops: the sides facing the wind had more colonies than the center or other edges of the crop. Thus gradients of virus disease parallel the activity of the vectors. When such gradients occur with persistent viruses, the vectors must have stayed in the area where they first landed; with nonpersistent ones, however, such gradients are more to be expected because vectors would lose their infectivity while feeding on plants near the edges of fields and would not infect plants at the center even if they later moved there. Trees, tall hedges, and buildings to the windward side shelter crops from aphids, but on the leeward side cause aphids to land, and lettuce mosaic is often more prevalent in parts upwind to such obstructions. Van der Plank (1948) pointed out the possible significance of crop perimeters in affecting the incidence of virus disease in crops covering different areas; the perimeter forms a greater proportion of a small than of a large field, and he reported that whereas maize streak virus often infects the whole crop in small fields, many plants in large fields escape infection.

The danger that a virus will be introduced into a crop is greater where insect host plants and virus sources are numerous than where they are few. Perennial plants are more dangerous than annuals, because once they are systemically infected, they usually remain potential sources of virus, but biennials can be almost as important as perennials in retaining virus from one year to another. Schlösser (1952) suggested that sugar beet viruses probably originated in the wild *Beta maritima* L., common on the coasts of Britain, and spread throughout Europe during the last 30 years. There can be little certainty about this kind of observation, however, because virus diseases are often overlooked until they are looked for critically.

Many annual weeds are potential sources of virus, but they are usually of little significance. Thus beet yellows virus infects *Chenopodium album* L. and C. *murale* L. in beet and spinach fields, but it rarely spreads from them to cultivated plants. However, several economically important viruses in the United States are carried to cultivated crops by leaf hoppers from weeds, some of which are annuals. Luicerñe dwarf virus, which also causes Pierce's disease of grapevines, can be transmitted from many species of naturally infected plants, and infectively leaf hoppers are found in such different habitats as cultivated valleys, high mountains, deserts, and seashores.

One of the most studied diseases is curly top of beet, transmitted to several crop plants in western United States by the leaf Hopper *Circulifer tenellus*, often during transient feeding when the insects move from overwintering hosts in the desert and foothills to cultivated valleys. Virus persists in some overwintering hoppers, and the insects breed on the virus-susceptible Russian thistle and wild mustard in the deserts during the summer and fall. Severin (1939) found 75 species of plants, several perennial that become naturally infected with curly top virus. Three perennials are food plants of the hoppers in uncultivated areas, and virus is carried from them to annuals which germinate after early rains. During 5 years with such rain up to 42% of the subsequent hoppers were infective, whereas during 2 years without early rain, the proportion was less than 6%.

A few aphid-transmitted viruses, too, seem to depend on weeds for their survival: celery yellow spot virus is not transmitted by mechanical inoculation or by several species of aphids from celery to celery, but *Rhopalosiphum conii* (Dvd.) (= *Hyadaphis xylostei* Schrank) from symptomless infected *Conium maculatum* L. transmit virus to celery and hemlock. Cereal yellow dwarf virus is transmitted by the 5 species of aphid that infest cereals in California. Rain delays the sowing of the cereals, but encourages the growth and subsequent heavy aphid infestation of grasses, many of which are susceptible to the virus. When drought follows, infective aphids move from the drying grasses into young grain fields. Simons et. al. (1956) described an interesting relationship between tomato and pepper crops and weeds infected with potato virus Y in Florida. Different strains of the virus occur in three widely separated areas, but not in two others only 50 miles away where suitable weed hosts and vectors are present. Potatoes were, or still are, grown in the affected areas, but not in the free ones, and as the distribution of diseased tomato and pepper crops bears no obvious relationship to potato crops, the authors suggest that the virus was introduced with potatoes and persisted in weeds.

Although it is realized that wild plants are often sources of virus from which epidemics may being, very little is known about the incidence of disease in them in most parts of the world. A few workers have started to survey the vegetation of prescribed area for virus diseases: MacClement and Richards (1956) in Canada testing with mechanical inoculation only, found about one plant in ten infected, many with viruses common in cultivated crops. This suggests that a large proportion of wild plants may be infected with one virus or another. In many areas of Britain, however, there is no evidence that susceptible weeds play a significant part in the epidemiology of common virus diseases of such crops as potatoes, brassicas, and lettuce. Infected tubers or seedlings, or plants in older crops are the main sources, and virus is spread from one crop to another when vectors seek alternative hosts. As they fly or are blown over a distance they tend to be dispersed, and the greater the distance between crops, the greater is the dispersion; consequently, crops near a virus source usually become more heavily infected than those farther away. Virus spread is greatly retarded, also, when susceptible crops are separated from one another by immune plants, especially if the intervening plants are suitable hosts for the vectors.

The economic importance of spread of virus from one crop to another depends largely upon the age and purpose of the healthy crop. Insects usually fly away from maturing crops, and if other susceptible crops in the area are at a similar stage and are to be harvested soon, infection will probably cause little loss. However, if young susceptible crops, plants for vegetative propagation, or biennials, for seed the next year are being grown, infection may have serious consequences. Spread of virus from one crop to another is particularly important in potatoes; in many parts of the world aphids disperse from them in midsummer, about 2 months before the crop is harvested. Other crops of the same age are visited and infected with virus even if not colonized by the aphids. The plants are usually too old to show symptoms, but seed tubers are infected and give poor yields the next year. In some countries a high proportion of the crops are infected, and so much virus is carried into new stocks that it is unprofitable to keep them for a second year. In Britain and the United States, however, most commercial potato crops are now fairly healthy, so virus spread from one crop to another is not great. Many horticultural crops such as lettuce are grown in small plots in Britain, and serious losses occur when lettuce mosaic virus is carried by aphids from maturing to young crops and there is no break in the succession of crops. The susceptible crop need not be colonized by the vectors, for aphids can infect most bean plants adjacent to clover fields with yellow bean mosaic although they rarely breed on them.

If biennials to be kept for seed become infected, they may from an important source of virus for the annual crop: a cycle of infection begins that can only be broken by growing the seed plants elsewhere, as was done with cabbage seed in the United States and cauliflower seed in England. Much work has been done on beet seed crops, for yellows virus not only halves the yield of sugar beet seed, but the plants can be a major source of virus for the root crops. Watson et. al. (1951) found that distance from a seed crop within a seed area has a pronounced effect on the incidence of mosaic in sugar beet crops but not of yellows: mosaic is usually confined to fields within 100 yd. of a seed crop. Virus is carried to the seed plants from the root crops during the autumn, and the vectors (*M. persicae*) also overwinter on them, carrying yellows to young root crops for miles around in the spring. Healthy seed crops are now produced in Britain by spraying stecklings with appropriate insecticides or raising them in cover crops or away from root crops.

F. Spread within Crops

Plants become infected and act as sources of inoculum within crops because (1) they grow from infected seed, (2) they grow from infected tubers or some other plant part either planted or remaining from a previous crop, (3) they become infected in the seedbed and are later transplanted, or (4) they are infected by incoming vectors. If virus is not brought into a crop from outside, the number of plants that becomes infected is often directly proportional to the number of initially infected plants, so the health of the crop at the beginning of the season is important.

Spread of viruses by insects within crops is usually over short distances, often to neighboring plants, more often along rows than across them, and sometimes in the direction of the prevailing wind. Spread often results in foci of infected plants around those initially infected, whether the virus is persistent or nonpersistent, or the vectors are aphids, beetles, or other insects. There has been much discussion about the relative importance of winged and wingless forms of aphids as vectors. Many workers have assumed that virus is spread from one crop to another by winged aphids, but that subsequent spread to nearby plants within the crop is by wingless ones e.g.. Direct observation of the movements of small insects is difficult, and has rarely been attempted. Those who watch flying aphilds record that they fly laterally from plant to plant, or over short distances, or they fly upward and are swept away by the wind. Others have shown that wingless aphids walk from plant to plant in potato crops, particularly when their leaves are in contact. Weather greatly affects the movement of wingless aphids, which move most often when it is hot, and especially when plants wilt.

Experiments on the time when viruses spread in potato crops show that much of the season's spread occurs early, when the colonizing winged aphids are active and before a wingless population develops. Doncaster and Gregory thought that wingless aphids might be responsible for the further spread of virus within the crop when the plants touch each other, because winged aphids rarely colonized potatoes during the summer dispersal. But it cannot be assumed that the winged ones do not visit potatoes because they do not colonize them, so they might also spread virus later in the season. The very significant correlation between trapped *M. persicae* and the spread of both leaf roll and Y viruses suggests that most spread is by winged forms. The lower correlation coefficient for rugose mosaic (Y) agrees with the evidence from Scandinavia that *M. persicae* is not the only vector of this virus.

Watson and Healy (1953) used statistical methods of relate trap catches or field counts of aphids to the spread of beet yellows and mosaic viruses in sugar beet crops, and concluded that winged *M. persicae* are most important in spreading beet yellows virus, despite the usual predominance of *aphis fabae* on the plants. It is probable that winged forms of both *M. persicae* and *A. fabae* spread beet mosaic virus from sources outside the crop, but little within it. Winged *A. fabae* are apparently not concerned in spreading yellows virus, presumably because they often remain on the first plant they colonize, whereas *M. persicae* moves from one plant to another for a few days, depositing nymphs in small batches.

Additional evidence that the winged forms are primarily responsible for spreading virus in potato crops was obtained by surrounding healthy plants with sticky boards to prevent aphids from walking away from adjacent infected plants. But some of the most conclusive evidence has come from experiments with insecticides. Emilsson and Castberg (1952) controlled aphids with parathion, but did not control the spread of potato Y virus, and Schepers and associates (1955) sprayed potatoes frequently with nicotine, preventing the development of any wingless aphids, yet there was considerable spread of both leaf roll and Y viruses, and the distribution of infected plants in sprayed and unsprayed plots was similar. Later trials with demeton, when no wingless forms developed, had little effect on virus Y, but the spread of leaf roll virus was greatly decreased. It was stopped, and that of Y virus decreased to about half with insecticides in Britain when virus was not introduced from outside the crop. Presumably aphids visit fewer plants in a crop treated with insecticide, and infect fewer with Y virus before they are killed; they die before becoming infective with leaf roll virus. Steudel and Heiling (1954) assumed that demeton affects the wingless forms only, and that much of the spread must be by wingless forms because spraying decreases the incidence of beet yellows. However, as most winged *M. persicae* visit several plants, spraying with a persistent insecticide will decrease the number visited and the incidence of yellows whether spread is by wingless or winged forms or both.

One of the reasons why some aphid species are important vectors whereas others, equally efficient in laboratory tests, are not, is that some lose their power of flight more readily than others. Young winged forms of some species are much more active than older ones because the wing muscles degenerate after the aphids find suitable hosts and start to reproduce. The more suitable the hot, the sooner the aphids settle and lose the power to fly, so aphids that are apparently well-adapted to their hosts, such as *Aphis fabae* on beet, are unlikely to be able to fly by the time they become infective with beet yellows whereas *M. persicae*, which does not colonize so readily, will still move occasionally from plant to plant. When aphids are newly mature, even host plants are visited and abandoned several times, and so good colonizers can be efficient vectors of nonpersistent viruses.

Although the evidence suggests that wingless aphids are of little importance as vectors, they do walk from one plant to another, and in hotter climates than northern Europe may move frequently and contribute largely to the spread of persistent viruses. Walking aphids might not be expected to transmit nonpersistent viruses readily because they are seldom infective after spending some hours undisturbed on an infected plant; however, many were infective after a short period of walking and probing on infected plants, and presumably those which walk off a plant have spent some time walking on it first. More information has been obtained on this by catching winged and wingless aphids soon after they voluntarily leave cauliflower plants infected with cabbage black ring spot virus of cauliflower mosaic virus and placing them singly on young seedlings. Similar proportions of winged and wingless aphids transmit virus. We cannot conclude, therefore, that wingless forms do not spread virus if they move, but only that they move infrequently in cool climates, as Fisken (1957) found in Scotland, and then perhaps inoculate adjacent plants, many of which have been infected already by winged forms.

Relatively little work has been done on the epidemiology of insect borne fungi or bacteria. Rankin et. al. (1941) surveyed 3000 square miles of New York state for Dutch elm disease and found at least 100,000 dead and dying trees. More than a third of *Scolytus multistriatus* beetles collected from elms carried *Ceratostomella ulmi*, but despite the prevalence of the pathogen and vectors, spread was slow, suggesting that the beetle is an inefficient vector or that other factors present limit infection. Two such factors are that beetles readily infect trees during June and early July, but usually fail to do so later, and infection often fails to become systemic. Local spread of the disease from isolated infected elms was also studied by Zentmyer et. al. (1944). Three-quarters of new infections occurred within 100 ft. of the source, and the maximum distance was presumed to be 180 ft., although the authors state that beetles sometimes carry fungus more than 2 miles. Spread was much more rapid than in New York, 40% of all trees within 75 ft. of the source becoming infected during 2 years. Statistical analysis showed that the probability that a tree would become infected decreased directly with the logarithm of the distance from the source; this result applies to the spread of most pathogens, whether insect- or air-borne. Zentmyer et. al. postulated that wind influences local spread as well as long distance because more trees were infected to leeward of the source of inoculum. The distribution of trees makes this argument extremely dubious, however, because there were several trees within 30 ft. of the source to leeward, none to windward.

Little is known about the movements of insect vectors within crops, except what can be postulated from the distribution of diseased plants. Direct studies of insect movement are likely now to become easier than they were, for we know more certainly what questions to ask, and can employ new techniques, such as marking insects or plants with radioactive isotopes, in answering them.

6

EPIDEMICS OF DISEASES

PART A
EPIDEMICS

This chapter is analytical rather than descriptive. This chapter aims at an understanding of the factors and processes that go to make an epidemic, and not at a description of important epidemics, past or present.

A. Some Common Misconceptions

It is stated in the literature that for an epidemic to occur there must be an aggressive parasite that multiplies fast, spreads far and swiftly, and is not particularly selective in its requirements. These are total misconceptions. They are the cause of much confused thinking and must be disposed of forthwith.

Consider swollen shoot of cacao, a systemic virus disease in West Africa, as an example. The virus complex is carried by slow-moving flightless mealy bugs of the family Pseudococcidae, and spreads largely from tree to tree in contact. It is delicate, and survives for less than an hour in the feeding vectors. The number of infected trees on a farm multiplies slowly; under conditions of unrestricted natural spread it took 30 months for the percentage of infected trees to increase from 31 to 75. The spread of infection from farm to farm is also slow. In 1947 the largest area of disease in West Africa had reached a radius of only some 10 miles after having spread continuously since 1922. Yet there can be no doubt that swollen shoot is a major epidemic. The virus complex is endemic in West Africa indigenous tree of the Sterculiaceae and Bombacaceae. As a sporadic disease, cacao swollen shoot has a long history in West African. But some 30 years ago the epidemic began, with the merging of separate but expanding, more or less circular outbreaks, into larger amorphous areas of dead and dying trees, which spread ever more rapidly and devastated whole farms in their spread. The disease has caused political upheavals, and the cacao industry from the western Ivory Coast to central Nigeria has needed drastic measures to save it.

Another example is oak wilt, caused by the fungus *Ceratocystis fagacearum*. The current epidemic of the disease in the United States was rated sufficiently high to get an entire chapter to itself in the limited space of "Plant Diseases," the United States Department of Agriculture's Yearbook for 1953. Yet the parasite has very limited means of spread. It can spread over a distance by some means that are not yet properly understood, but the main spread is from tree to tree. Infection takes place through root grafts, and local barriers such as roads are sometimes enough to stop it.

An extreme example is psorosis, a virus disease of citrus. No vector is known. The virus can spread from tree to tree by root grafts, but the natural spread is so slow that it is extremely difficult to demonstrate in orchards. Yet psorosis is currently the greatest killer of citrus trees in California. Psorosis epidemics, it seems, may take half a century or more to develop.

B. Epidemics as a Matter of Balance between Opposing Processes

The human population of a country can increase as a result of a high birth rate; it can also increase as the result of a low death rate. It will increase with a low birth rate provided that the death rate is even lower. As with men, so it is with plant pathogens. Disease can increase to epidemic levels because (to consider only the extremes) the pathogen has a high birth rate or because it has a low death rate. The misconception we have just discussed have arisen because attention has been focused on high birth rate epidemics. But low death rate epidemics are also important, particularly with perennial hosts.

High birth epidemics are usually caused by fungi which develop uredospores, conidia, oidia, and the like, and produce local lesions in the host. Low death rate epidemics are caused largely but not entirely by systemic pathogens. The explanation is often fairly obvious. The systemic pathogen is safe within the host, and if the host is perennial and the disease not immediately lethal, a long life of the infected host guarantees a long life to the systemic pathogen. With pathogens as well as with men longevity means a low death rate. The terms are practically equivalent.

The line between the two types of epidemic is often somewhat blurred, but at the extremes the distinction between them is clear and has interesting consequences. High birth rate epidemics are usually controlled by fungicides or resistant varieties; low death rate epidemics are mainly controlled by means of sanitation. High birth rate epidemics usually spread fast; low death rate epidemics usually spread slowly. These and other consequences will unfold themselves logically as we proceed. But the terms high birth rate epidemics and low death rate epidemics have been used purely to illustrate a point and now will be dropped in favor of more conventional expressions.

I. THE MULTIPLICATION OF INFECTIONS

A. Multiplication within a Crop

1. *The Course of an Epidemic*. There is no such thing as a typical epidemic. The variety of epidemics in plant pathology is infinite. To mention just two factors, the epidemic varies with the amount of inoculum that is the source of the infection, and this ranges with the different diseases from scarce to abundant; it varies with the rate at which the infections multiply, and this ranges from very slow to very fast. Because of this variety, one example is as good as another. Blight of potatoes, caused by *Phytophthora infestans*, is chosen here as an example because of its place in the history of plant pathology. But even for this one disease the example we have chosen for illustration is not typical of all epidemics. It is based on data from western Europe. If blight in North America had been the example, stress would have been laid on the danger of epidemics starting from piles of culled potatoes and garbage dumps near towns.

Figure 6.1 is based on data taken from a probit foliage decay line given by Large (1945) for an epidemic of blight on potatoes in England. Observations were started on August 11, when 0.1% of the foliage was infected. This is equivalent to an average of about 1 lesion per plant. The epidemic progressed for about 4 weeks, after which all the foliage had been destroyed by blight. The curve that shows this in Fig. 6.1 is sigmoid. The figure also shows the rate of increase of infection per cent per day. The rate for this epidemic began at about 48%, when observations were started on August 11, and gradually dropped to zero as the epidemic ran its course.

In the Netherlands the early history of a potato blight epidemic was studied by Van der Zaag (1956). We assume that the results hold for England too Van der Zaag found that the most important primary foci of infection were infected plants growing from infected seed. These plants had a few weak, shriveled shoots with a few lesions from which spores were released during the second half of May. They were most abundant in the very susceptible variety, Duke of York, in which about 1 primary focus per square kilometer was found. Infection spread first around these foci, then generally over the fields, and from field to field and from variety to variety. Our concern here is not with the progress in any particular field but with the epidemic generally.

These findings show how inadequate a picture Fig. 6.1 gives of an epidemic. There are 3,000,000 or 4,000,000 plants in 1 sq. km. of potatoes, so that there are (in round figures) 3,000,000 or 4,000,000 lesions per square kilometer of potatoes when 0.1% of the foliage is infected. Still in round figures, infection had to multiply 1,000,000 times from the original foci before the level of 0.1% was reached on August 11. Fig. 6.1 describes the one-thousandfold increase (from 0.1 to 100%) that occurred after August 11, and ignores the one-millionfold increase that went before. And in terms of time Fig. 6.1 starts in August instead of May, like the biography of a centenarian that starts after his seventieth birthday.

Consider the matter from the farmer's point of view. For him the most important single characteristic of an epidemic is its date of onset. This decides when he must spray or whether he need spray at all. If the date is early—say, in July in England—heavy losses are likely to occur if the disease is unchecked. But if it is late—say, in September—losses are likely to be small (apart from tuber infection), and spraying would be a waste of material and effort. There are of course some variations in detail (an epidemic may start early and then be checked by a change in the weather) but the general pattern is clear, as can be seen from the analyses of Moore (1943), Large (1952), Beaumont et. al. (1953), and Large et. al. (1954). The date of onset is determined by what happens before the onset, i.e., by that portion of the epidemic not described in Fig. 6.1.

Infection increased one-millionfold between the second half of May and August 11, i.e., in about 80 days. This is equivalent to an average increase "at compound interest" at the rate of 17% per day. But the rate would not have remained at a steady average since it is affected by changes in the weather. Apart from that, the potato becomes more sus-

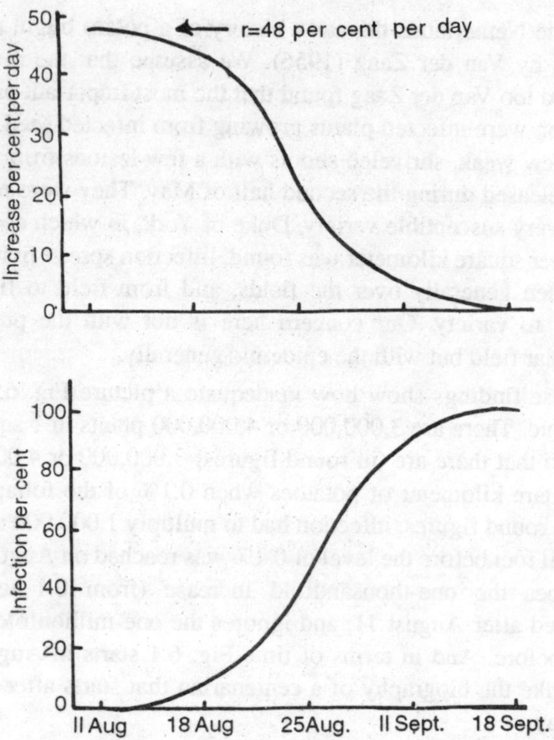

Fig. 6.1. *Lower half*: The progress of an epidemic of blight (*P. infestans*) in potatoes. (Data of Large 1945). *Upper half:* The increase of infection per day expressed as a percentage of the infection already present.

ceptible as it becomes older. One can therefore infer that the rate was considerably less than 17% at the start and increased to 48% by August 11.

2. *Rate of Multiplication before the Onset of the Epidemic*. For convenience we can define the onset of an epidemic as that time when not more than 5% of the susceptible tissue is infected, or, for systemic diseases, when not more than 5% of the plants are infected. Within this rough limit we can decide on any criterion we wish; for example, with potato in England and Wales the starting point for the assessment of blight is 0.1% infection, and this level can conveniently be taken to indicate the onset.

The rate of increase of disease has been taken as

$$\frac{dx}{dt} = kx\,(1-x) \tag{1}$$

in which k is a constant and x is the proportion of susceptible tissue that is infected, or, for systemic diseases, the proportion of infected plants. The rate at which infection occurs is, according to this equation, proportional to the product of the proportion of infected tissue and the proportion of healthy susceptible tissue, that is, to the proportion of tissue that is still available for infection. Curves of this equation describing the progress of infection with time have the sigmoid shape so commonly found in plant pathology. Large (1945) applied the equation to blight epidemics and concluded that it would give a good fit to actual blight progress curves within the limits of accuracy of the original blight observations. Nevertheless the reader is warned that after the onset a good fit can come about only as a result of counterbalancing factors, because the incubation period (i.e., the pre-reproduction period after infection) inevitably tends to increase the value of k as the epidemic progresses beyond the onset. That is, after the onset, k is just an empirical constant even in a constant environment. We use equation 1 and sigmoid progress curves almost solely to provide a helpful introduction to the study of multiplication. After the introduction the topic is treated in a way that will nearly obviate the need for considering disease progress curves beyond the onset.

Infection is influenced by the weather, the susceptibility of the host, the virulence of the pathogen, etc. So equation 1 will be rewritten in the form

$$\frac{dx}{dt} = kx\,(1-x)\,f(T,\ H,\ S\ldots) \tag{2}$$

in which $f\,(T,\ H,\ S\ldots)$ is a function of the temperature and humidity of the air, the susceptibility of the host, and so on.

Before the onset of an epidemic, x is small and, in the definition already given for the onset, less than 0.05. As a good approximation for the progress of infection up to the onset, the term $(1-x)$ can be dropped and equation 2 rewritten as

$$\frac{dx}{dt} = kxf\,(T,\ H,\ S\ldots) \tag{3}$$

$$r = \frac{100}{x}\frac{dx}{dt}$$

$$= 100kf\,(T,\ H,\ S\ldots) \tag{4}$$

By definition, r is the rate of increase per cent per day (or per year, if years are the units of time) up to the onset of the epidemic. Within the limits of the approximation just stated r is independent of the progress of the epidemic (up to the onset) and becomes a direct reflection of the environmental conditions, the susceptibility of the host, and so on. A qualification, normally of minor importance, will be discussed in Section II, A, 6.

3. *Some Comments about r*. One must become familiar with r. It is a fundamental and perhaps the most useful concept of epidemiology. It plays its part not only in multiplication rates, but also in the rate of spread of epidemics in distance. (Section III, A and E), in the theory of forecasting (Section II, B), in sanitation (Section II, C), and directly or indirectly, in much else. The impression may have been given that r is only an approximation. That is incorrect. For convenience r is allowed to make its debut approximately as the percentage rate of multiplication at or before the onset of an epidemic, which is accurate enough for present purposes. But r is a concept capable of exact determination (as by equation 5) and applicable throughout the course of the epidemic. The reasons for using it mainly at or before the onset will appear later.

In reporting the results of censuses, human birth rates, death rates, and rates of increase of population are given in terms of 1,000 of population. It is not assumed that the rates are the same for all hours of the day, for all seasons of the year, or for all years of boom and depression, peace and war. The crude rates are simply a convenient way of expressing what is happening. So too r can be regarded as a convenient way of reporting the results of censuses of infections before or at the onset. It is not implied that r is constant from hour to hour or day to day. It just expresses the tempo of developments up to the onset in an uncomplicated way. But one condition is implied: fractions of a day or, if appropriate, a year must be avoided. When, for example, it is calculated in Section II, C that an epidemic would be delayed 3.7 days this must be interpreted as "from 3 to 4 days."

In one way r has an advantage over crude census rates, which are subject to the influence of varying proportions of the population in the different age groups, particularly to a varying proportion of women of childbearing age. Superficially, the same problem occurs with r. With potato blight, for example, it takes from 4 to 6 days or longer after inoculation for spores to be produced. But for most of this period there is no recognizable lesion, and spore formation begins soon after the lesions become visible in the field. Up to the time of the onset of the epidemic relatively few of the lesions seen in the field are not of spore bearing age; one could without great difficulty count as lesions only those of an age to produce spores. The validity of using lesions of this sort for determining multiplication rates is examined in Section II, A, 6.

The magnitude of r reflects the effect of temperature, humidity, rainfall, wind, and other environmental factors on multiplication; it reflects among other factors the resistance or susceptibility of the host, the proportion of spores that germinate, the proportion of germinated spores that manage to enter the host and establish infection, the rate of increase in size of the lesion, the speed at which spores are produced and their number, the abundance of vectors and their mobility, efficiency, and distribution, and the size and number per acre of the host plants. It summarizes the equilibrium of the factors (except x itself) that affect the tempo of multiplication. Because of this, few of these factors will be specifically dealt with in this chapter. There is, for example, no discussion of the effect of weather and climate. This may seem strange in a chapter on epidemiology. But it is logical. When a forecaster uses his knowledge of the weather and other relevant factors to forecast an epidemic, he is in fact forecasting a faster rate of multiplication. To him the weather and other factors are important. To us (for the purpose of this chapter) the rate of multiplication is important. This statement carries no implication that the one approach is better than the other. There are several possible approaches to the discussion of epidemics. One chooses what one believes to be the most apt for one's purpose.

4. *The Estimation of* r. Substitute $r/100$ for $kf(T, H, S...)$ in equation 2, integrate, change logarithms to the base 10, and rearrange thus:

$$r = \frac{230}{t_2 - t_1} \log \frac{x_2 (1 - x_1)}{x_1 (1 - x_2)} \tag{5}$$

Here x_1 and x_2 are the proportion of susceptible tissue infected (or, for systemic diseases, the proportion of infected plants) at dates t_1 and t_2, and $t_2 - t_1$ is the interval in days (or years, if this is appropriate) between these dates.

For example, according to the data of Large (1945) blight in potatoes at Dartington in 1942 increased from 0.1 to 5.0% in 7 days. Here x_1 and x_2 are 0.001 and 0.05, respectively, and $t_2 - t_1$ is 7 days. From equation 5, $r = 57\%$ per day.

In general, the use of equation 5 assumes that the lesions are randomly distributed. In the particular case of potato blight there is the independent evidence of Gregory (1948) that this is so; but randomness is not the rule. There are often quick departures from randomness, especially when systemic diseases multiply, particularly the systemic diseases of trees. There is a general rule about this: the smaller the size of lesion, the higher the value of x_2 that can be used without incurring a serious error from lack of randomness. A systemically infected plant must be considered as a single lesion (Section III, F).

As another example, consider cauliflower mosaic. It spreads in non-random fashion to form nests of infected plants. In a trial with different strains of cauliflower there were 1.7 and 5.9% infected plants on August 21 and September 4, respectively (Fig. 6.4)), so x_1 and x_2 are 0.017 and 0.59, respectively, and $t_2 - t_1$ is 14 days; whence r is 9.2% per day. The epidemic was started artificially by planting an infected plant at the center of each plot of 121 plants, so 5.9% infection represents roughly 7 plants around each infector plant. The evidence is that at this stage the error from nonrandomness is small, so the estimate of r is fairly trustworthy; but at much higher values of x_2 estimates would be too low.

With local lesion disease on plants that grow between t_1 and t_2, allowance must be made for new susceptible tissue. If susceptible tissue increases m times between t_1 and t_2 equation 5 must be rewritten as

$$r = \frac{230}{t_2 - t_1} \log \frac{m x_2 (1 - x_1)}{x_1 (1 - x_2)} \tag{6}$$

Chester (1943) gives figures for the multiplication of leaf rust (*Puccinia triticina*) of wheat in Oklahoma. After a severe winter followed by normal weather, infection increases at a steady rate from 1 pustule per 1,000 leaves at the beginning of March to 10 pustules per leaf at the beginning of June. During this period there is a tenfold increase of leaf tissue; hence m is 10. The ratio x_2/x_1 is 10,000, and $t_2 - t_1$ is 92 days. An infection of 10 pustules per leaf is equivalent to about 1% infection, so x_2 is 0.01 and x_1 is even smaller, and we can ignore the terms $(1 - x_1)$ and $(1 - x_2)$. By equation 6, $r = 12.5\%$ per day.

5. *General Comments on the Restriction of Much of This Chapter to Events before the Onset.* Physical chemistry advanced far in the theory of dilute solutions. The gas laws were applied to dilute solutions, and

dissociation constants were determined in dilute solutions. This was done in the first place to avoid the difficulties of dealing in theory and practice with the behavior of molecules and ions at high concentrations. In the same way this chapter deals to a great extent with the theory of dilute concentrations of disease: of disease up to the onset, defined arbitrarily as the 5% level. Primarily the reason for this is to avoid the objection against using disease progress curves after the onset.

How much information do we lose by confining quantitative discussions to the period up to the onset? The answer is: surprisingly little. Consider these examples. In Section II C an equation is given that evaluates sanitation in terms of a delay in the onset of the epidemic. If the weather and other conditions stay constant, the delay in the onset is also the delay in reaching the 50% level of disease or the 99% level. If the weather changes, then interest is changed from the factor of sanitation to the factor of weather. By describing the effect of sanitation on the onset, one has in effect described all that needs to be described about the effect of sanitation on the course of the epidemic. In other words, the loss of information from confining attention to the period up to the onset is negligible. Also, it does not matter what criterion one takes for the onset, provided that it is not more than about 5%; it could just as well or better be taken at 0.1%. In Section II B we discuss the application of *r* to the theory of forecasting; in most cases forecasts are in practice confined to the onset of an epidemic.

Gradients of infection away from the source figure largely in Sections III and V. With gradients, too, it can be shown both that it is wise to keep calculations to the period up to the onset of the epidemic and that remarkably little information is lost by doing so. In this chapter we do in fact discuss gradients at higher levels of disease, and correct the data as best we can. But this is only through lack of choice; the data are so scant that one cannot at present afford to be selective.

6. *Justification for Using the Law of Compound Interest when there is an Incubation Period.* The meaning given to *r* is that it is the rate of compound interest (per cent) up to the onset. True compound interest (we are not concerned with bankers' compound interest) is added to accumulated capital as it is earned, and then instantly begins earning itself; the concept of an incubation (i.e., pre-reproduction) period is foreign to the law of compound interest as it is used in science. Are we then justified in applying the law, as represented by *r* to plant disease?

It seems that we are, and it will be shown that r can be used as an apparent rate.

The meaning of x is the proportion of infected tissue in which incubation is complete. The infection is visible and the lesions are of an age to produce inoculum (according to the suggestion in Section I, A, 3). Consider now the total proportion x' of tissue that is infected, even if incubation is still incomplete. To simplify the argument, we shall confine attention to events before the onset of the epidemic. Instead of equations 3 and 4, let us write for the same set of observations.

$$\frac{dx'}{dt} = \frac{r'}{100} x'_{t-P} \tag{7}$$

Here x'_{t-P} is the proportion of infected tissue at time $t - P$, in which P is the incubation period. It is the proportion of tissue in which incubation is complete and therefore has the same meaning as x in the earlier equations. It can be determined that

$$r = r' e^{-\frac{Pr}{100}} \tag{8}$$

In effect, r is a compound interest rate to which P contributes, outwardly like any other factor, such as the abundance of vectors.

Equation 8 holds only when P and r' are constant. More generally, it can be shown that despite an incubation period the concept of compound interest is valid even if P and r' vary with temperature, humidity, etc.

There is more to it than just that. In the early history of an epidemic (for example, soon after inoculation in an artificial epidemic) r at first varies with time, even if P and r' are constant, and only later settles down to the stable value given by equation 8. The larger the product Pr, the larger are the variations. Similar variations occur when, for example, an epidemic is checked by drought and then builds up again. The previous history of the epidemic, every previous fluctuation, is remembered in the multiplication rate, and this memory factor is the special contribution of an incubation period to the concept of compound interest.

There is a qualification. Previous memory is wiped out each time multiplication stops for a continuous interval at least as long as P, and then starts anew. Hence, the memory factor can cause a drift in the multiplication rate of potato blight from day to day, but not in that of potato leaf roll from year to year.

Variations in the value of r, whether caused by memory or any other factor, are properly cared for by equation 5. This equation estimates the average of the value of r at every instant between any times t_1 and t_2, irrespective of any variations that occur.

But although it is important to know that the memory factor's effect or r is correctly estimated, it is often equally important to know how to eliminate this effect so that other factors can be studied without interference. Except within approximately 2.2 P days after inoculation in an artificial epidemic or 1.2 P days after the end of a major interruption in a natural epidemic, the effect of the memory factor on r can be made small by arranging that the interval between t_1 and t_2 should be as near to 1.2 P days (or a multiple thereof) as is possible without using fractions of a day. For example, if P is 6 days and it is wished to study in an artificial epidemic the separate effect on r of factors other than the memory factor, observations should begin whenever convenient after the thirteenth day from inoculation and then be made at weekly intervals. Very exact knowledge of P is here unnecessary, and in the example just quoted an 8-day interval between t_1 and t_2 would also almost eliminate the memory factor's effect on r.

B. Increase of Disease when the Pathogen Does Not Spread between the Host Plants

Consider bunt of wheat caused by *Tilletia caries* and *T. foetida*. Plants are infected as young seedlings, and do not release spores until the grain has been formed. One plant cannot infect another during the course of the season. There is multiplication of infection but only as an increase from season to season and not within a single crop. Many other systemic smut fungi behave similarly.

With obligate heteroecism there are similar restrictions. The apple rust fungi *Gymnosporangium* spp. move hither and thither from apples to the alternate hosts and back. During this movement from one host to the other there can be a multiplication of infection, but there is no multiplication directly from apple to apple. Similarly, *Cronartium ribicola* does not spread from pine to pine.

With many of the systemic or quasi systemic vascular wilt diseases the pathogen is not returned in great quantity to the soil until the host dies. The pathogen may build up from season to season, but it does not spread much from plant to plant during the season.

Sometimes there is apparently no spread from plant to plant and no building-up within the host species even over the seasons. Eastern X-virus spreads from chokecherry (*Prunus virginiana*) to peach but apparently not from peach to peach. The virus of Pierce's disease has a wide range of species of host plants from which it infects grapevines, but it does not seem to spread from grapevine to grapevine. Tomato spotted wilt virus infects tobacco, but the thrips vectors do not breed on this host and there is no evidence of spread from tobacco to tobacco. Peaches with X disease, grapevines with Pierce's disease, and tobacco with tomato spotted wilt apparently do not contribute to the building-up of an epidemic of these three virus diseases; in them epidemics are secondary (Section II D) and are the result of multiplication in other hosts.

Absence of spread from plant to plant can occur at times even with diseases that normally spread. Figure 6.2A, reproduced from a report by Doncaster and Gregory (1948), shows the distribution of rugose mosaic, caused by virus Y, in an isolated potato field. This field was initially free from virus Y, but was invaded from a source 300 yd. away. The plants which were infected as a result of this invasion did not pass the infection on to their neighbors, and there was no evidence of secondary spread. Absence of secondary spread can be expected with the invasion takes place late in the season or when the presence of vectors is transient. It is possibly not uncommon.

The type of increase without multiplication which has been discussed in the previous five paragraphs can cause an epidemic which superficially resembles other epidemics but which is fundamentally distinct. There is, for example, no reason why the progress curve of the epidemic with time should be sigmoid. Often the approach to control is different. It is therefore important to be able to detect when infection is increasing without spreading from plant to plant.

This can be done in various ways. One can mark infected plants and determine whether they are foci of infection by comparing the number of plants which become infected in an area near them with the number of plants which become infected in a comparable area at a distance away from them. But the most convenient methods use the distribution of diseased plants in the field or orchard without a knowledge of their history. Figure 6.2B shows how leaf roll develops in a potato field around an infected plant grown from an infected tuber. There is a nest of diseased plants near this plant. Nests of this sort vary in their compactness and shape from disease to disease, but it is probably safe to assume that with all plant disease there is a tendency for infected plants to aggregate to some extent near the source of infection. (The tendency is greatest when r' is low in value and the gradients are steep; see Section III E and F

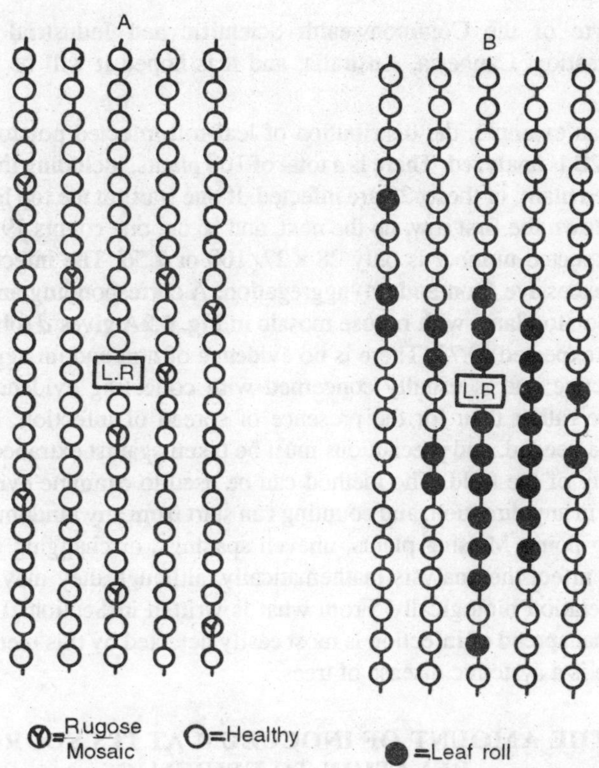

Fig. 6.2. Distribution of rugose mosaic and leaf roll around an infected plant, L.R., grown from a potato tuber infected with leaf roll. For convenience, rugose mosaic and leaf roll are shown in separate diagrams, A and B, respectively.

about scales of distance). The test for spread of infection is a test for aggregations. Cochran (1936) examined some general tests, and Todd (1940) and Freeman (1953) developed a test for the special case of the death of trees planted at the corners of a square lattice in a rectangular plantation. The simplest test yet devised makes use of doublets. A doublet consists of 2 adjacent diseased plants. If n plants are examined in sequence, and of these μ are diseased, the expected number of doublets is

$$d = \frac{1}{n}\mu(\mu - 1). \tag{9}$$

A run of 3 diseased plants is counted as 2 doublets; a run of 4, as 3 doublets; and so on. A test of significance has been worked out by G.A.

McIntyre of the Commonwealth Scientific and Industrial Research Organisation, Canberra, Australia, and it is hoped it will be published shortly.

As an example, the distribution of leaf-roll-infected potato plants in Fig. 6.2B is analyzed. There is a total of 100 plants, including the original diseased plant; of these, 28 are infected. If one starts at the top left corner, reads down the first row, up the next, and so on, one counts 19 doublets. The expected number is only $28 \times 27/100$ or 7.56. The infected plants show excessive (nonrandom) aggregation. A corresponding analysis for the 9 potato plants with rugose mosaic in Fig. 6.2A gives *d* (observed) 0 and *d* (expected) 0.72. There is no evidence of nonrandom aggregation. In practice one is usually concerned with collecting evidence for the absence rather than for the presence of spread of infection. Extensive data are needed, and precautions must be taken against extraneous heterogeneity of the field. The method can be used to examine evidence for spread in any direction, and counting can start from any randomly chosen starting point. Missing plants, uneven spacings, or changing directions do not affect the analysis mathematically, although they may affect its interpretation biologically. From what is written in Section III F it follows that spread of infection is most easily detected by this method if the disease is a systemic disease of trees.

II. THE AMOUNT OF INOCULUM AT ITS SOURCE IN RELATION TO EPIDEMICS

A. The Amount of Inoculum and Rate of Multiplication before the Onset

In the preceding section our concern was with rates of multiplication. The amount of inoculum at the source of the epidemic must now be considered. When infected plants are the starting inoculum—when, for example, epidemics of potato blight start from lesions on infected shoots developing from infected tubers or when potato leaf roll spreads from an infected plant, the sequence of multiplication is homogeneous. Lesions produce lesions, and infected plants produce infected plants. We shall be concerned only with homogeneous sequences. When the source of inoculum is resting spores or anything but infected plants, we consider the sequence as starting from the first plants to be infected directly from this inoculum, i.e., the first infected plants are considered as the inoculum at its source. This restriction to homogeneous sequences simplifies what has to be discussed without affecting the conclusions appreciably.

Suppose I_0 and I are respectively the amount of inoculum (in terms of the number of lesions or, for systemic diseases, the number of infected plants) at the source and at the onset of the epidemic which was defined in Section I, A, 2.

$$I = I_0 e^{rt/100}. \tag{10}$$

Here t is the time taken up to the onset, and r has the meaning given in earlier equations. To quote an example given previously, if *P. infestans* multiplies a millionfold in 80 days.

$$e^{80r/100} = I/I_0 = 10^6$$

whence $r = 17.3\%$ per day. This is an average rate for the 80 days.

More often one needs to know the effect of a change in the amount of inoculum at the source: a change brought about, for example, by sanitation. One can therefore profitably rewrite equation 10. Suppose the amount of inoculum at the source is reduced from I_0 to I'_0. The delay in the onset of the epidemic, Δt, can then be determined from the equation

$$I_0 = I'_0 e^{r\,\Delta t/100}$$

whence

$$\Delta t = \frac{230}{r} \log \frac{I_0}{I'_0} \tag{11}$$

In the equation, r is the rate during the time of the delay. Logarithms are to the base 10.

Suppose $r = 69.3\%$ per day at the onset. How long will the onset be delayed by halving the amount of infection is doubled every day, halving the amount of inoculum at its source?

$$t = \frac{230 \log 2}{69.3}$$

$$= \frac{230 \times 0.301}{69.3}$$

$$= 1.0.$$

The answer is 1 day. Natural compound interest of 69.3% per day, which is added to the accumulated capital at every instant, is equal to 100% "banker's" compound interest, which is added once a day. This numerical problem makes the obvious point that if the amount at its source will postpone the onset by a day. The criterion for the onset must of course be fixed, but, provided that it is fixed within the limit stated in Section II,

A, 2, the particular value at which it is fixed has no relevance. (But read also Section II, A, 5).

Consider another example. Broadbent (1957) tested the use of barriers to protect cauliflower seedlings from mosaic, and found that a barrier of three rows of barley around the seedbed reduced infection from 3.0 to 0.6%. How long would be onset of an epidemic of cauliflower mosaic be delayed in crops planted from protected seedbeds? Use the value $r = 9.2\%$ per day, calculated for cauliflower mosaic in Section II, A, 5.

$$\Delta t = \frac{230}{9.2} \log \frac{3.0}{0.6}$$

$$= 17.5$$

The onset of the epidemic would be delayed 17.5 days. For properly apt results, r should be measured for the same variety as that used in the barrier experiment, in the same district, at the same time of the year, in the same type of soil, etc.

B. Some Comments on the Theory of Forecasting of Epidemics

The method and effectiveness of forecasting epidemics varies with the value of r in equation 11. Consider the two extremes of forecasts that ignore, and forecasts based entirely on, the amount of inoculum at the source.

1. *Forecasts That Ignore the Amount of Inoculum.* According to equation 11, the higher the value of r at the onset, the less the error of a forecast of the time of onset that ignores differences in the amount of inoculum at the source. If r is large enough, approximate forecasts become possible without reference to inoculum. An alternative condition would be that the amount of inoculum is so large that it ceases to be a limiting factor. Both conditions possibly exist with, say, apple scab when scabby spring weather follows a fall and a winter favorable to the survival of large amounts of inoculum in dead leaves.

Potato plants become more susceptible to blight as they grow older, and high values of r are possible toward maturity. It is this fact of a high value toward maturity that makes short-range forecasts of blight feasible on weather data alone. It has been found that there may be a "zero time", a date (July 1 in west Scotland) before which forecasts based on weather data alone are invalid (Grainger, 1950). Interpreted by equation 11, "zero time" is the date on which the magnitude of r becomes high enough to blur the effect of variations in the amount of inoculum and make the epidemic specially sensitive to the weather. The more susceptible the variety, the earlier can "zero time" be fixed or, alternatively, if "zero time" is unchanged, the more accurately can forecasts be made.

2. *Forecasts Based on the Amount of Inoculum Alone.* As the value of r at the onset decreases, the effect of variations in the amount of inoculum on the date of the onset increases. With low enough values of r the importance of the inoculum factor becomes dominant, and approximate forecasts can be based on it alone. The number of sclerotia of *Sclerotium rolfsii* in the soil can be used to forecast losses of sugar beet from this fungus. In general the soil is not a medium conducive to the fast multiplication of pathogens, and methods of forecasting other diseases caused by soil-borne pathogens on inoculum alone are likely to be feasible.

C. An Epidemiological View of the Problem of Sanitation: the First Rule of Sanitation

Equation 11 tells us that, measured as a delay in the onset of an epidemic, the benefit of a given percentage reduction in the amount of inoculum at its source is inversely proportional to r, the magnitude of r being determined at the time of the delay. This is the first rule of sanitation.

Within the limits of one disease the rule means that sanitation helps most when it is needed least. Before any method of sanitation is recommended for the control of a disease, it should be ascertained that the method will remain effective even during seasons when conditions are such that infection multiplies at its fastest.

In comparing different diseases the rule means that control by sanitation is most apt for those disease that multiply most slowly. If r is large, sanitation is relatively ineffective, and it is usually necessary to use fungicides or, alternatively, to call in the plant breeder to produce a resistant variety, i.e., to bring down the magnitude of r. When r is small, one is likely to be dealing with a disease that can be controlled by sanitation: by crop rotation, by fumigation of the soil before the crop is planted, by destroying diseased crop residues or using manure free from inoculum, by planting healthy or disinfected seed and nursery stock, by roguing out diseased plants, by isolating fields from sources of inoculum, by destroying weeds or other hosts that can carry infection, or by any other method that reduces inoculum at its source.

For example, systemic diseases tend to have low values of r, and a study of the literature shows that the majority of them are controlled by sanitation in one form or other.

As r increases, sanitation becomes less effective. But where do we draw the line? At the one extreme, with very low rates of multiplication, sanitation is almost completely effective; for example, on present knowledge, a citrus grower can be protected against losses from psorosis by the sanitary measure of using only healthy nursery stock to plant his orchards. Are there, at the other extreme, infections that multiply so fast that sanitation is worthless? The answer depends on the circumstances. Consider this example. A field of potatoes grown from blight free seed can become infected with blight from infected refuse piles or from infected fields. Take the second alternative. From calculations based on meager data in the literature, it can be estimated that doubling the isolation of a potato field from its neighbors would reduce the amount of blight that the field gets from its neighbors by 83%. (See Section III B and D). If, from Fig. 6.1, we take r to be 48% per day, this isolation will postpone the onset of an epidemic by 3.7 days. (Here $I_0/I'_0 = 100/(100-83)$). On an allowance of an increase of a ton of tubers per acre per week during the critical period of the growing season, isolation to the extent stated would mean a gain of crop in an unsprayed field of about half a ton per acre. Whether this gain would be worth working for depends on the circumstances. On a small farm, isolation is impossible or possible only at the cost of arrangements that are not acceptable to the farmer. But on a large estate some farm planning with an eye to isolation is not ruled out. It is not for our present purpose desirable to pursue this topic at length, but it should be stated that even with a fast multiplying disease like blight of potatoes the possibility of sanitation cannot be excluded before conditions are analyzed.

D. An Epidemiological View of Breeding for Resistance

One especially important deduction from equation 11—and the first rule of sanitation—concerns resistant varieties. Plant breeders like to aim at high resistance or complete immunity; this is desirable, and the effect of immunity is readily understandable: there is no epidemic (and in that case the matter would be outside the scope of this chapter). But commonly the available resistance is only partial; infection occurs, but at a slower rate. The magnitude of r is reduced, and the value of sanitation correspondingly increased. Every quantum of resistance, however small, increases the efficiency of sanitation; and every quantum of sanitation increases the value of the partial resistance. As methods of control, sanitation and partial resistance should go together. When a disease is controlled by sanitation, partial resistance in any measure is an achievement not to be despised. It has been much undervalued in plant pathology, largely as the result of ignoring its connection with sanitation.

E. Secondary Epidemics

An epidemic can be defined as secondary if it starts from inoculum derived from another, earlier epidemic. In Section I B it was observed that epidemics of eastern X disease of peaches, Pierce's disease of grapevines, and tomato spotted wilt on tobacco are secondary because there is apparently no spread of the diseases in the crops mentioned. Infection must come from outside.

In the Netherland van der Zaag (1956) found that with the potato variety Duke of York there was about 1 primary focus of infection per square kilometer at the start of the season, this focus originating from an infected seed tuber. With the more resistant variety Noordeling, primary foci were much rarer or entirely absent. Further, he tested varietal differences by artificially inoculating leaflets in the field before natural infection was apparent, and then 12 days later counting how many leaflets had become secondarily and naturally infected from these primary artificial sources. For every 1 lesion on Noordeling that developed secondarily around the primary source, there were 180 on Duke of York. From this we infer that the magnitude of r was roughly 180 times as great for Duke of York as for Noordeling. This estimate is rough but adequate; the evidence does not allow an exact estimate of either r or r'. Epidemics in Duke of York take about a month to develop from the primary foci, so, from equation 11, an epidemic in Noordeling would need 180 months to develop. If one adds to this the information about the scarcity of natural primary foci with Noordeling, the period needed would be much greater. Even if one admits the possibility of error, the differences are striking enough to make it reasonably certain that a primary epidemic of blight in Noordeling is improbable in a growing season limited to 4 or 5 months. If Noordeling becomes significantly blighted, it is almost certainly as a result of infection obtained initially from some other variety.

The tendency for disease to spread from more to less susceptible varieties and cause secondary epidemics in them, increases with the extent to which initial inoculum must multiply to cause an epidemic. For any given time up to the onset, the logarithm of the amount of multiplication is proportional to r. If two varieties differ in that in the one r is twice as great as in the other at all times, then by the time the more susceptible variety has multiplied 10^2 times the less susceptible will have multiplied 10 times; by the time the more susceptible has multiplied 10^6 times the less susceptible will have multiplied 10^3 times. As multiplication continues, the difference in level of disease between the varieties increases. The more the multiplication needed before a level of, say, 5% is reached, the greater is the difference in disease between the varieties when the more susceptible has reached that level; hence, other things being equal, the greater the chance that disease in the less susceptible variety will be influenced by inoculum from the more susceptible. With diseases like potato blight, in which relatively little inoculum survives the winter, the danger of susceptible varieties starting secondary epidemics in other varieties is correspondingly great.

Late blight, caused by *Phytophthora infestans*, can survive perennially on tomatoes in countries in which tomatoes are grown all the year round. It can also be brought into the summer-grown tomato plants of colder eliminates by blight-infected transplants imported from warmer areas. But late blight of tomatoes has also a considerable literature of epidemics secondary to those of potatoes. Neighboring blighted potato fields or potato refuse piles are repeatedly implicated as the source of tomato blight. But in one detail there has been a noteworthy change in the literature since Mills (1940) summarized it. In 1940 races of *Phytophthora infestans* from potato were poorly adapted to tomatoes, and according to Mills, had to be trained to attack tomatoes by a few passages through tomato leaves. Nowadays there are frequent references in the literature to potato races that readily attack tomatoes, although a distinction between potato and tomato races still exists. On the general evidence of the literature, secondary epidemics on tomatoes are now more easily established from potatoes. There has been, it would appear, a change in the relationship between *P. infestans* and the potato, which has also involved the tomato secondarily and possibly incidentally, and which has perhaps been the cause of the destructive epidemics of tomato blight on a scale unknown before 1946.

There is a type of secondary epidemic brought about the mixed cropping. The apple variety Delicious has considerable resistance to powdery mildew, caused by *Podosphaera leucotricha*. Jonathon is very susceptible and is commonly used as a pollinator for Delicious. In mixed orchards spores from diseased Jonathans may start secondary epidemics on Delicious which are severe enough to require control measures. Hardwoods grown alone are normally resistant to *Fomes annosus* but may be killed when grown in mixed stands with susceptible conifers in which inoculum builds up. These examples show the value of uniformity of resistance in minimizing the risk of secondary epidemics, a point to remember when reading Sections V, C, 3, f and V, E.

Secondary epidemics provoked by mixed cropping are often used by plant breeders to eliminate susceptible lines of seedlings. A variety known to be susceptible is interplanted with the material to be sorted out in order to ensure that infection will be present in the breeders' plots. The method is simple and convenient, but there is the danger of confusing the smaller amount of resistance adequate to stop a primary epidemic with the larger amount needed to cope with a secondary epidemic.

III. THE SPREAD OF EPIDEMICS

A. Factors That Affect the Rate of Spread

Section III describes how an epidemic spreads over a distance. Three factors are considered at the start. A fourth factor, scale of distance, is discussed later in the section.

1. *The Gradient of Infection.* The probability that a healthy plant in a given direction will become infected depends on the distance of that plant from the source of infection. Infection grades away (usually smoothly, but not necessarily so) from the source. Unless otherwise stated, the source will always be taken to be a point—a single plant, for instance—and not a field or strip. Gradients should be determined only when the percentage of infection is low; apparent gradients necessarily become flatter as disease mounts.

2. *The Abundance and Distribution of Susceptible Host Plants.* The probability that a spore or other propagule will travel a given distance is proportional to the number of propagules released at the source. If the number of air-borne spores at the source is multiplied 100 times, then on an average 100 times as many as before will blow past the first milestone, 100 times past the second, and so on. When heavy spore showers are detected hundreds of miles from the nearest source, it is reasonably certain that the source is not a single plant but fields of plants, not just a few fields but hundreds of acres of fields. Abundant host plants mean, too, that the leaps the pathogen need make from one host plant to another are small. Their distribution—whether in small fields or large—

also affects the pathogen's spread; this is a matter to be discussed in Section IV.

3. *The Rate of Multiplication.* Spread is multiplication at a distance from the source of inoculum; and, for a given gradient, multiplication at a distance is related to multiplication in general. Fast multiplication of disease lesions means fast spread of disease. The connection has long been tacitly accepted, and pathologists commonly use the words "multiplication" and "spread" interchangeably. A tie between multiplication and spread is discussed in Section III, E (equation 13).

B. Horizons of Infection

Horizons of infection have been discussed by van der Plank (1949). For argument's sake suppose temporarily that the probability that a lesion (or, for systemic diseases, an infected plant) will form a daughter lesion in unit time on a healthy plant at a distance s is

$$p = \frac{a}{3^b} \tag{12}$$

Here a and b are constants. We shall not bother about the accuracy of this equation; the relationship shown is just scaffolding, which will be removed later. Suppose that there are infected fields scattered over a large area, and consider any field as center. Suppose that this field is only lightly infected: that the epidemic in it has not gone further than the onset, and that it receives inoculum from other fields in a sector narrow enough for the gradients toward the center to be considered uniform. The probability that a lesion will develop in this field as a result of inoculum received from a parent lesion two units of distance away in the sector is $1/2^b$ times the probability that a lesion will develop from inoculum from a parent lesion one unit of distance away in the sector. But on an average the number of parent lesions two units of distance away is twice the number one unit away (because for a given angle at the center the arc is proportional to its radius). Hence on an average the number of lesions caused by inoculum received from all sources (all parent lesions) two units away is $2/2^b$ or $1/2^b$ times the number from inoculum received from all sources one unit away. Similarly the number from inoculum received from all sources three units away is $1/3^{b-1}$ times the number from all sources one unit away. If Q is the number of lesions from inoculum received from all sources one unit of distance away the number of lesions from inoculum received from all sources at all distances is

$$Q\left(1+\frac{1}{2^{b-1}}+\frac{1}{3^{b-1}}+\ldots\right)$$

If $b > 2$, this series is convergent, and there will be a limit from beyond which inoculum will not come.

To simplify calculations, suppose that the distance between fields is relatively large and that the fields are uniformly infected and uniformly distributed. We can, as an adequate approximation, take the distance between the centers of neighboring fields as the unit of distance. If $b = 2.5$, 62% of the daughter lesions caused by inoculum received from other fields will come from fields, beyond the immediate neighbors, 40% from more than 3 fields away, and 32% from more than 5 fields away. If $b = 3$, the respective figures are 39, 17, and 11%; if $b = 4$, the figures are now 17, 3, and 1% respectively. If, for example, one regards inoculum coming from behind the horizon as negligible if it is responsible for less than 10% of the daughter lesions, then if $b = 2.5$, a horizon is established more than 50 fields away; if $b = 3$, about 5 fields away; if $b = 4$, 2 fields away. The horizon draws in sharply as b increases, i.e., as gradients become steeper.

It is not assumed that gradients are the same in all sectors, and the horizon about a field need not be circular.

C. Continuous and Discontinuous Spread of Epidemics

If $b > 2$, the source of a new outbreak will probably be within a horizon. The greater the magnitude of b, the more likely is the source to be near, and the easier it will be to follow the path of an epidemic. The spread will be continuous.

But if $b < 2$, more inoculum will arrive from far than from near (assuming of course that host plants occur over a wide area). Infection will appear as if "from nowhere." There will be what have been called "spot" infections—infections that cannot be traced to their source. The spread will be discontinuous.

Low value of b can be expected if the movement of inoculum is oriented with a restriction on random scattering, as would occur, for example, if the inoculum were carried by birds migrating toward a particular destination. Migratory birds are thought to spread chestnut blight by carrying the sticky pycnospores of *Endothia parasitica*. "Spot" infections were a feature of the great chestnut blight epidemic in North America, and occurred in addition to local infections caused by wind-blown ascospores; it seems likely that during migrations b was less than 2.

D. The Determination of Gradients; the Scrapping of Equation 12; Dutch Elm Disease; Potato Blight

Equation 12 was used only to build up an argument. In scrapping it one need not replace it by a better equation, but only consider its most useful features. The central inferences from the equation concern the value $b = 2$, or, to scrap the equation, concern a gradient in which the number of lesions formed by inoculum from a source varies inversely as the square of the distance from the source. If one plots the number of lesions against distance on a log-log scale, the inverse-square lines are straight. One can draw as many as one wishes—all parallel. Such are the lines *A, B, C, D* in Fig. 6.3. The observed gradient for any disease can then be compared immediately with these lines. If its slope is steeper than theirs, there will be a horizon of infection, and spread will be continuous. Similarly one can draw a number of parallel inverse-cube lines like the lines *E* and *F* in Fig. 6.3. If the observed gradient is steeper than theirs, horizons can be expected to be fairly close. In Fig. 6.4 the process is continued, and lines *G, H, I*, and *J* show the number of lesions inversely as the fourth power of the distance from the source of infection.

Dutch elm disease, caused by *Ceratostomella ulmi*, was chosen to illustrate Fig. 6.3. There is a fair amount of information about it. Among other things, the spread of disease from a single infected tree in a limited period of time has been observed by several workers to virtually cease within some hundreds of yards from the tree, so one knows in advance that one should expect gradients steeper than the inverse-square lines. On one point there is a difficulty—a difficulty common to most of the literature of diseases that could be used for illustration: gradients should be determined only at low percentages of disease. This limits information to the lower part of the curve; the disadvantage of this is that information there is usually based on relatively few diseased plants and is consequently not powerful statistically. At higher levels of disease one can correct the curve partially (or fully, if the lesions are distributed randomly) by transforming percentages of diseases into calculated numbers of infections per 100 plants. Attention to this was drawn by Gregory (1948), who published a useful table of transformed values. Some examples will explain the transformation. Suppose there were exactly 100 random infections per 100 plants. Not all plants would be infected; on an average, 36.8% would have no infections (they would remain healthy)

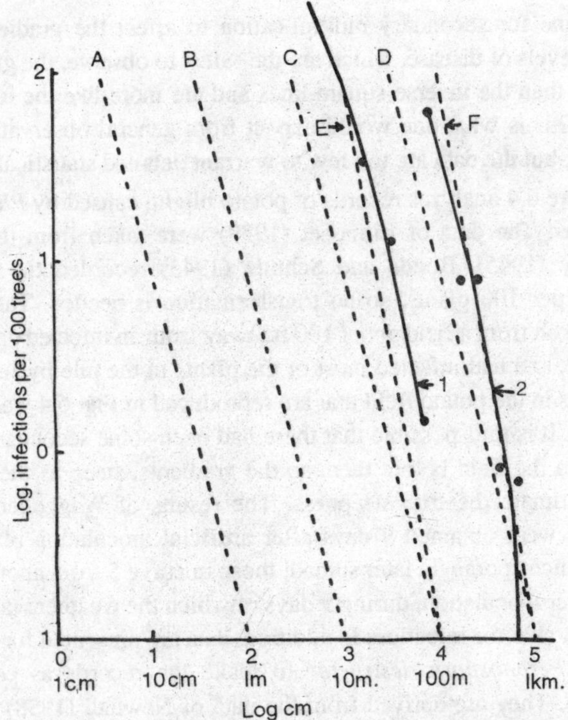

Fig. 6.3. Amount of Dutch elm disease (*Ceratostomella ulmi*) at varying distances from the source of inoculum. Curves 1 and 2: data of Zentmyer et. al. (1944) and Liming et. al. (1951), respectively, shown as transformed numbers of infections. A, B, C and D are inverse square lines with scales of distances as 1 : 10 : 100 : 1000. E and F are inverse-cube lines with scales of distance in the ratio 1 : 10.

and 63.2% would have 1 or more infections. The table calculates this in reverse; it transforms 63.2% disease to 100 infections per 100 plants. In Fig. 6.3 the highest figure for disease is 88.9%, which is transformed to 220 infections per 100 trees. At low percentages the change is small and usually negligible: 5% disease transforms to 5.13 infections per 100 plants.

Figure 6.3 analyzes the combined data of Zentmyer et. al. (1944) for three plots in Connecticut; and the data of Liming et. al. (1951) for a plot in New Jersey. These data were collected within a year of the emergence in large numbers of the vector beetles from the source of infection which in each case was a single naturally infected tree. There was therefore little time for secondary multiplication to affect the gradients. At the lower levels of disease, which are the safest to observe, the gradients are steeper than the inverse-square lines and are more like the inverse-cube lines. This is what one would expect from general observations on the disease, but the data are too few to warrant detailed statistical analysis.

Figure 6.4 analyzes records of potato blight, caused by *Phytophthora infestans*. The data of Limasset (1939) were taken from the paper of Gregory (1945). Bonde and Schultz (1943) recorded the number of lesions per 100 plants, so no transformation is needed. These records were taken from a field about 100 ft. away from an infected dump pile on June 12, and had infected most of the plants in the pile by June 25. The readings in the potato field that are reproduced in Fig. 6.4 were taken on July 12. It is thus possible that there had been some secondary multiplication in the field before then, so the gradients, steep as they are, may underestimate the true steepness. The results of Waggoner (1952) in curve 4 were obtained 9 days after artificial inoculation of the center from which sporangia later spread; those in curve 5 from another plot 18 days after inoculation, during 9 days of which the weather was generally unfavourable for infection. In addition, data are presented for the related fungus *Peronospora destructor* to make the records as complete as possible. They are derived from the data of Newhall (1938) for downy mildew lesions per 100 ft of row, and no transformation was needed. These various records are consistent in suggesting that, as a reasonable approximation at the low amounts of disease in which we are primarily interested, the number of lesions falls off inversely at least as the fouth power of the distance from the source. The result has been assumed in the calculation made for blight in Section II, C. In Fig. 6.5 (to be discussed later) two more curves are shown for *Phytophthora infestans*. They are slightly steeper than an inverse—sixth-power gradient. The evidence from all the records that the curves at low amounts of disease become steeper than an inverse—fouth-power gradient is statistically significant; but it is inadequate to say how steep they eventually become.

Curve 3 in Fig. 6.4 is displaced to the right of curves 4 to 5. Displacements will be discussed under the next heading. But one reason for displacement seems evident here: the source of infection for curve 3—the cull pile—was comparatively large and heavily infected, and was at a distance from the field; the sources for curves 4 and 5 were smaller and within the fields.

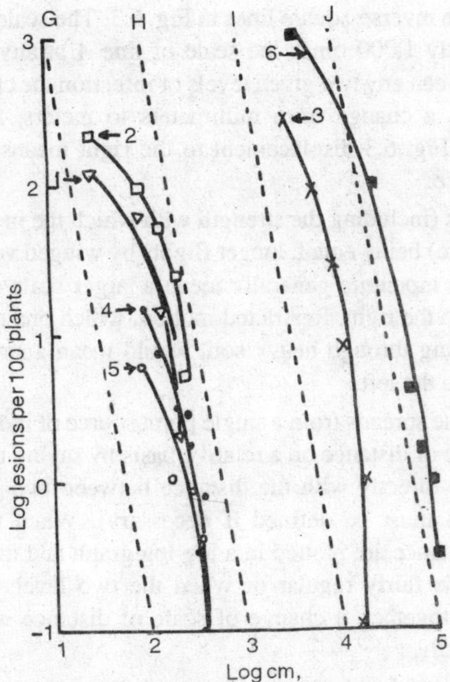

Fig. 6.4. Amont of potato blight (*P. infestans*) at varying distances from the source of inoculum. Curves 1 and 2: data of Limasset (1939). Curve 3: data of Bonde and Schultz (1943). Curves 4 and 5: data of Waggoner (1952). Curve 6: data of Newhall (1938) for *Peronospora destructor*. curves 1, 2, 4, and 5 are based on trasformed number of lesions. G, H, I, and J are inverse fourth-power lines with scales of distance as 1 : 10 : 100 : 1000.

E. Scales of Distance; a Tie between Multiplication and Spread

For illustration let us return to the chestnut blight fungus spread by migrating birds. There are two relevant features about birds during migration: the flights are longer between pauses, and are oriented toward some particular destination. The second feature—orientation—was cited as probably involved in making spread discontinuous. The first feature—long flights—was ignored. Making flights longer without change of orientation and without serious loss of inoculum during flight would make an epidemic spread on a larger scale (in the word's literal meaning of relative dimensions). But it would not make a continuous spread discontinuous.

Consider the inverse-square lines in Fig. 6.3. The scale of distance for line *D* is exactly 1,000 times the scale of line *A* at any given level of infection. Between any two given levels of infection the change from line *A* to line *D* is a change from millimeters to meters, from meters to kilometers. In Fig. 6.3 displacement to the right means an increase of scale of distance.

Other things (including the strength with which the inoculum is emitted at the source) being equal, longer flights by winged vectors or longer motions by the inoculum generally mean a larger scale of distance and displacement to the right. Restricted motion, which one might expect of inoculum moving through heavy soil, would mean a smaller scale and displacement to the left.

When disease spreads from a single point source of inoculum, one can define the scale of distance on a relative basis by saying that the scale of distance varies directly with the distance between two given levels of disease (which must be defined if necessary). When curves relating disease and distance are plotted in a log-log graph and the displacement of the curves is fairly regular or when the two levels of disease are selected close together, a change of scale of distance is readily determined graphically.

Consider Fig. 6.5, which is constructed from data of van der Zaag (1956) for the spread of *Phytophthora infestans* from an incompletely

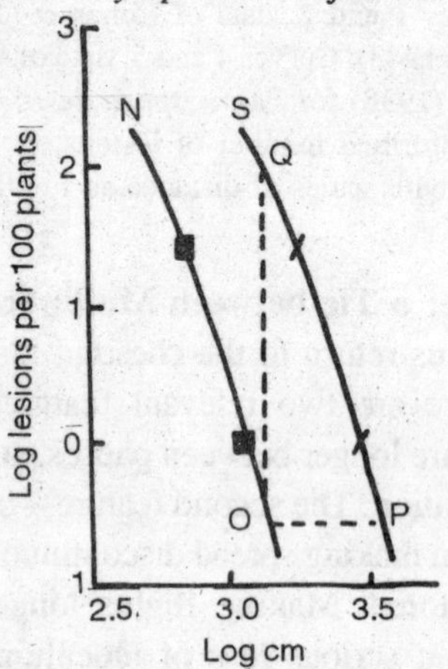

Fig. 6.5. Amount of potato blight, *P. infestans*, at different distances from the source of inoculum. Data of van der Zaag (1956), given as transformed numbers. Explanation in text.

removed source of infection. The data for percentages of diseased plants have been transformed into the number of lesions per 100 plants. To the north of the focus (curve *N*) the spread was less than to the south (Curve *S*). Since the spread in both directions was from the same source of inoculum, the displacement of curve *S* to the right of curve *N* is the result of a changed scale of distance. This is measured by the line *OP*, which represents an increment of 0.4 in the logarithm of the distance or a 2.5-fold increase in scale of distance at the level of disease at or near points *O* and *P* (log 2.5 = 0.4). The argument will have been grasped in the simpler comparison between the straight lines *A* and *D* a few paragraphs back, and needs no further elucidation.

The corresponding change in the scale of disease at the distance represented by points *O* and *Q* is measured by the line *OQ*, which represents an increment of 2.5 in the logarithm of the number of lesions or a 316-fold increase in the number (log 316 = 2.5). At this distance there was about 316 times as much disease to the south as to the north as a result of a 2.5-fold change of scale of distance.

The records of Waggoner (1952) for *P. infestans* in his plot at Clear Lake in 1950 on day eighteen show a scale of distance to the NNW about 8 times as great as to the SSE. Since, approximately, disease varied inversely as the fourth power of distance from the source (curve 5 of Fig. 6.4), there was about 8^4 or 4096 times as much disease over a given distance to the NNW as to the SSE. This result appears to have been due to wind. There is considerable evidence on the literature that wind can strongly alter the scale of distance. One reason why wind affects the scale in different directions is that more inoculum leaves the source in one direction than in another.

To determine a relation between multiplication and spread, referred to in Section III, A, 3, consider a primary gradient set up in healthy plants surrounding a point source of infection. A primary gradient is one set up by inoculum derived directly from the initial point source without complications from secondary multiplication of disease in the zone of the gradient. For example, curves 1 and 2 of Fig. 6.3 probably represent primary gradients of Dutch elm disease. Return to equation 7 and consider a boundary condition. While $t < P$, the quantity x'_{t-P}, taken here as the point source, is constant. Because the gradient is set up in plants healthy at the start, i.e., when $t = 0$, the amount of disease at any given time less than *P* and (for a given gradient) at any given distance from the source, will be proportional to the value of r' prevailing over the period of the observations. Whence, to follow the reasoning in the previous paragraphs,

$$\Delta s = kr'^{(1/b)} \tag{13}$$

where Δs, which determines the scale of distance, is the distance between two given levels of disease in the gradient, k is a constant; b has the same meaning as in equation 12; and changes in r' are assumed to be moderately small.

If the two levels of disease are taken fairly close together, the value of b can, with reasonable accuracy, be taken as constant over Δs. This feature appears in all the curves in Figs. 6.3 and 6.4, especially at distances not too near the source. Therefore the use of b here does not revive equation 12, but only the part of it that may legitimately by revived. It is necessary that b should be positive; i.e., there must be an actual decrease of disease with distance away from the source.

Equation 13 holds for those factors (except the incubation period) which effect r but not the gradient. With such a range of factors affecting r—from wind to the susceptibility of the host—and such a range of inoculum—from air-borne spores to water-splashed bacteria and vector-transmitted viruses—any generalization about the effect on gradients is difficult. It is, however, worth noting that in their effect on scale of distance the factors that determines r cannot be distinguished one from another (expect the incubation period). Consider, for example, diseases that spread mainly "by contact" between plants that are immediate neighbors. It is common experience that these diseases multiply slowly, and it is generally assumed that they multiply slowly because inoculum can only be spread by contact. In many cases one can use the opposite argument: the diseases appear to spread by contact because they multiply slowly for reasons that need not be known.

A summary of the basic concepts might facilitate an understanding of the relation between multiplication and spread. We have been concerned with a point source of inoculum, hence only with primary gradients, hence not with the incubation period, and hence with r' and r, as equation 13 shows. If inoculum comes from a source other than a point (e.g., an epidemic's front) the incubation period and r may become involved. A particular solution for the relation between r and the spread of the front of an epidemic of wheat stem rust underlies the Appendix. It is for the special case in which the pathogen's birth rate greatly exceeds its death rate and lesions are small.

F. Inverse Scales of Distance; the Law of Lesion Size

We have been considering changes in the distance scale on which a pathogen can spread. Now let us think of the problem from the opposite view ad consider changes in the scale of distance over which a pathogen must spread if it is to transmit disease. As an example we may consider the law of lesion size.

The law is that the highest potential rates of multiplication are with small lesions. A systemically infected plant is regarded as a single lesion extending over the whole plant. (The largest lesion is a systemically infected tree). In round figures, small local lesions like those of potato blight can multiply one-billionfold in a season, systemic diseases of small herbaceous plants up to ten-thousandfold, and systemic diseases of trees about tenfold in a year. Evidence is given elsewhere. The word "law" is used to indicate that the relationship was observed, not predicted. Only maximum rates are relevant; it is not relevant to know that in arid climates or immune varieties potato blight does not multiply at all.

Two virus diseases with efficient aphid vectors are cauliflower mosaic and tristeza disease. The number of cauliflower plants infected with cauliflower mosaic multiplies faster than that of orange trees infected with tristeza disease. Suppose cauliflower plants in a field could be enlarged to 10 times their normal size with all distances increased in proportion: the plants would be 10 times as tall, 10 times as broad, with the distance between them 10 times as great, and in rows 10 times as wide. Suppose one plant in the field is infected. The distance from any relative position on this plant to any relative position on a neighboring plant or any relative position on any plant in the field would be increased 10 times, by the process of enlargement. The scale of distance that an aphid would have to travel to transmit the virus would be increased 10 times, and the probability of a transmission would drop, correspondingly, according to the gradient. Similarly, if orange trees in an orchard could be shrunk to one-tenth their size with all other distances shrunk proportionately, the probability of transmission would rise in accordance with a decrease to one-tenth of the scale of distance over which an aphid must travel. So, other things being equal, an aphid as a greater chance of transmitting mosaic between cauliflower plants than between orange trees simply because orange trees are larger. The argument applies not only to vector-transmitted viruses but to any sort of inoculum. It applies not only to systemic disease but to any lesion; a uredospore of a cereal rust fungus must travel over only a very small distance, a few millimeters, to fall clear of an isolated pustule on healthy tissue. Cereal rust pustules can multiply fast because the scale of distance the pathogen must travel is small.

Inverse scales of distance are only one of the factors involved in the effect of size of lesion. The incubation period (to mention another) is often longer in systemic disease; the effect of this period has already been given in equation 8. Forest trees seem to have resistance to virus disease, systemic bacterial diseases and systemic smut diseases. Operating in the opposite direction is the factor of the size of plant (Section VI, D, I).

IV. EPIDEMICS IN RELATION OF THE ABUNDANCE AND DISTRIBUTION OF HOST PLANTS

A. General

Agriculture has meant bringing plants together in fields and—to meet the ever growing demand for food—increasing the acreage of these fields. This raises two separate questions with which this section is concerned.

First: how does the bringing together of plants and increasing the acreage affect the prevalence of disease? The general answer is that disease is encouraged. But there are exceptions, and increasing the acreage, if it means enlarging the fields without increasing their number, may at times even reduce the percentage of disease. Also, apart from this exception, the relation between disease and intensified agriculture is not simple. Some diseases are more sensitive to intensified agriculture than others, and increase relatively faster as the acreage is increased.

The second question is: how does the distribution of host plants affect epidemics? For example, if the acreage remains constant, how do the size and distribution of the fields influence infection? The answers can be of the greatest practical importance if epidemics can be reduced by nothing else than planned agriculture. The least exploited method of reducing plant disease is by planning the pattern of farming in directions other than by crop rotation; no attempt is made here to produce a blue-print for such exploitation, but it is hoped at least to indicate that blue-prints are entirely feasible.

B. Disease in Susceptible Plant Scattered among Immune Plants

Although bringing plants together into fields normally increases the likelihood of disease, epidemics among scattered host plants are not uncommon. For example, *Puccinia malvacearum* attacks only wild and garden plants. Yet it spread through western Europe within 5 years of its first appearance.

But bringing plants together may sometimes necessary condition for an epidemic to start. A good example is provided by the "native" potatoes of the mountains of Basutoland. The chief vector of the potato viruses, *Myzus persicae*, is abundant, and its winter host, the peach, almost ubiquitous. The varieties have no great resistance to virus disease, and succumb quickly when they are planted in garden plots. But when they grow—as they ordinarily do—as odd plants in the corn fields or in the grass near roadsides, there is little evidence of virus disease, and the varieties have persisted for more than a century in good health.

Mixed cropping has been deliberately used to control disease. Planting squashes and marrows in corn is commonly practiced in South Africa to control an aphid-borne mosaic disease. In Denmark and England beet seedlings intended for seed crops have successfully been protected against virus yellows disease by shield crops of barley. In France the idea has been extended. Seedlings of beets are raised in beds which are latticed with longitudinal and cross bands of barrier crops to give subdivisions about 20 yd. long by 5 yd. wide. The barrier bands are about 1 yd. wide. Oats, corn, hemp, and sunflower are generally used. In England, Broadbent (1957) showed that narrow barriers of barley rows reduced the incidence of cauliflower mosaic and cabbage black ring spot in cauliflower seedbeds, a matter that has already been mentioned in Section II, A as an illustration of a quantitative problem in sanitation. There is a considerable range in detail as one passes from mixed cropping through shield crops to barrier crops; but all are methods based on the reduction of the effective movement of inoculum, and all share the common feature that they are useful in practice only against a type of disease which we shall call a crowd disease, i.e., a disease that is likely to reach epidemic proportions only when the host plants are crowded together into fields.

C. The "Epidemic Point" in Crowding Plants Together

Before crowd diseases are discussed consider what happens when susceptible host plants are crowded together. It is useful here to return to our introductory concept of epidemics as a matter of balance between the birth and death rates of the pathogen. Bringing plants nearer together increases the movement of inoculum between them; and it increases the birth rate of the pathogen. But it is unlikely to affect the death rate greatly (unless the pathogen or its vector is itself parasitized or preyed upon).

When the point is reached at which births exceed deaths the danger of an epidemic beings.

Take as examples three disease of trees: rosette disease of peaches, swollen shoot of cacao, and tristeza disease of sweet oranges on sour orange rootstocks. All are caused by viruses that kill their hosts and die with them. Rosette virus ordinarily spreads slowly from peach to peach, and its birth rate is low. However, its death rate is high, because it quickly kills its peach host, which usually dies in the same season in which symptoms are first seen, and then dies too. Ordinarily, the death rate exceeds the birth rate and the disease in peaches is self-eradicating although it occasionally flares up and affects a whole orchard. Swollen shoot of cacao has the same features as peach rosette, but not in quite such an extreme degree. It spreads slowly, but not quite so slowly; it kills quickly, but not quite so quickly, and infected trees may survive up to 2 years or longer. Tristeza disease has an abundant and efficient vector, *Toxoptera citricidus* (among others), which occurs in both winged and wingless forms, and infection spreads relatively fast through an orchard in which *T. citricidus* occurs. The disease kills sweet oranges on intolerant rootstocks but not very fast; the trees go into a decline that may last for several years. Compared, then, with peach rosette, citrus tristeza has a higher birth rate and a lower death rate.

The tree diseases from an interesting series with regard to the epidemic point, which we define as the point in crowding plants together at which the birth rate of the pathogen exceeds the death rate.[*] In peach rosette the epidemic point is usually imaginary: it is usually not reached even when peaches are in continuous orchard formation. The disease of peaches is usually sporadic and self-eliminating, and the virus persists only because it has other more tolerant hosts such as the wild plum, from which it spreads to peaches. If epidemiology is the science of disease in populations, in peaches this behavior is epidemiological hypersensitivity, comparable in its effect on populations with ordinary hypersensitivity in its effect within the plant. Swollen shoot of cacao has a lower epidemic point, which was apparently reached in West Africa between the World Wars, when cacao production was expanding rapidly. Before then the disease in cacao had a long history of sporadic outbreaks; it was only afterwards that the epidemic began to rage. Tristeza seems to have a

[*] This definition is loose because the balance between births and deaths varies with x. A precise definition would need to stipulate some arbitrary value of x at which births should still exceed deaths.

fairly low epidemic point in countries in which *Toxoptera citricidus* is present; in these countries the virus seems to pervade quickly all susceptible species of citrus, killing those trees which are on unsuitable rootstocks and saturating the rest.

It is reasonable to suppose that with all diseases that cause primary epidemics the host plants are crowded beyond the epidemic point, at least during epidemic years. (The epidemic point will necessarily be lower in years favorable to the disease). If the epidemic point is passed during the recorded history of man's crowding plants together, there will be a discontinuous change from sporadic disease to epidemic disease, as has happened apparently with swollen shoot of cacao. But where the epidemic point is low it was probably passed long ago, and the process of intensifying agriculture will seem to have a continuous effect—with epidemics becoming continuously more likely to occur.

D. Crowd Diseases; Changes in the Relative Importance of Disease; a Glimpse of the Future

Crowd diseases are likely to reach epidemic proportions only when the host plants are crowded together; they are diseases with a high epidemic point. The inoculum is likely to have a high death rate (as when, for example, it cannot persist in the soil); or the movement of inoculum between separated plants is likely to be small because of a small scale of distance, as reflected in a steep gradient and a low value of r or, more strictly, r'.

By equation 11 the effect of variations in the amount of inoculum reaching a field is greatest when r is least. To make a long story short, consider only diseases like potato blight, the cereal rusts, and those virus diseases not transmitted directly through the soil or dead plant residues. Usually, the systemic virus diseases multiply more lowly than the local lesion diseases. Therefore, intensified agriculture, by bringing fields nearer together and increasing the movement of inoculum between them, is likely to advance epidemics of systemic virus diseases relatively faster than those of local lesion diseases. To some extent the trend has been observable during the past century of fast rising food production. Plant pathology started as a branch of mycology, but became increasingly more concerned with virus diseases. One grants that much of this history has been fortuitous; had de Bary's chief interest been in virus diseases, the history would have been different. But the change was not wholly fortuitous.

Further, because diseases especially sensitive to the intensification of agriculture are, as a result of this characteristic, especially amenable to control by isolation and sanitation generally, plant pathologists can expect to be much more concerned with the theory and practice of sanitation in the future than in the past.

E. The Effect of Making Fields Larger

1. *Inoculum Coming from Other Hosts.* Consider the finding of Wellman (1937) in Florida that cucumber mosaic virus, causing mosaic in celery, comes from other hosts such as *Commelina nudiflora* growing within 75 ft. or of vectors flying right over the field without settling in it. The area of 75 ft. of the field. For the sake of argument, accept this figure and ignore the possibility of some inoculum arriving from beyond 75 ft zone about a square field of 0.1 acre is 0.86 acre; about a field of 1 acre it is 1.85 acres; about a field of 10 acres it, it 4.95 acres. Thus 10 acres of celery divided into 100 square fields of 0.1 acre, would draw infection from other hosts scattered over 86 acres; divided into 10 square fields of 1 acre, from other hosts scattered over 18.5 acres; and concentrated into 1 square field of 10 acres, from other hosts scattered over only 4.95 acres. Concentration of the host plants into a single field of compact shape reduces infection from outside to a minimum.

Probably one of the few occasions on which the percentage of disease is reduced while agriculture is intensified is when infection comes from scattered hosts near the fields and acreage is increased by increasing the size of the fields and not their number.

2. *Inoculum Moving between Fields.* We shall consider here what happen when fields are made larger and correspondingly fewer, so that the total acreage remains constant. To do so we shall revive equation 12, but only to the limited extent that we use it, as in Section III, D, to illustrate the gradient of disease. Discussion will be limited to what happens before the onset of a epidemic, i.e., to inoculum moving into fields that are not already heavily infected, which is the only case of practical importance.

Assume that the fields are equal in size and uniformly shaped, orientated, and distributed over the country. Making them larger and correspondingly fewer increases the average distance between them in proportion to the square root of their average area, their shape and orientation being unchanged. The average distance, not necessarily on a straight course, that a spore (or other propagule) must travel from any particular relative position in the field of its origin to any particular relative position in the first field it reaches is proportional to the square root of the average area of the fields, that is, it must travel an average distance $k_1A^{1/2}$ where k_1 is a proportionality constant and A the average area of the fields. The probability that a lesion (or, with systemic diseases, an infected plant) in one field will form a daughter lesion (or infected plant) at any particular point in the first field is, therefore, by equation 12

$$\frac{a}{k_1^b A^{1/2\,b}}$$

The average distance the spore must travel to cross the first field, that is, the average potential number of daughter lesions along its track across the field, is also proportional to the square root of the average area. The probability of a daughter lesion somewhere along the track can therefore be taken as

$$\frac{ak}{k_1^b A^{1/2\,(b-1)}}$$

where k is another proportionality constant. Similarly, if the spore passes the first field, the probability of a daughter lesion in the second field it reaches is

$$\frac{ak}{k_2^b A^{1/2\,(b-1)}}$$

in the third

$$\frac{ak}{k_3^b A^{1/2\,(b-1)}}$$

and so on where k_2, k_3...... are constants. The probability of a daughter lesion in any field other than the field of origin is:

$$p = \frac{ak}{A^{1/2\,(b-1)}}\left(\frac{1}{k_1^b} + \frac{1}{k_2^b} + \frac{1}{k_3^b} + \ldots\ldots\right)$$

The series

$$\frac{1}{k_1^b} + \frac{1}{k_2^b} + \frac{1}{k_3^b} + \ldots\ldots$$

may be taken as convergent when $b > 1$ because it is almost identical with the series

$$\frac{1}{k_1^b}\left(1 + \frac{1}{2^b} + \frac{1}{3^b} + \ldots\ldots\right)$$

which is convergent when $b > 1$. Hence, when constants are collected together as K,

$$p = \frac{aK}{A^{\frac{1}{2}(b-1)}} \qquad (b > 1) \tag{14}$$

within the limit of the initial assumption of uniformity.

As an example of the effect of area, with an inverse-fourth power gradient (which on present data seems a fair approximation to the behavior of *Phytophthora infestans* not too near its source), doubling the average area of fields and halving their number reduces the probability that a lesion will cause the development of a daughter lesion in some other field by 65%; trebling their area reduces the probability by 81%.

If one wishes to show that the number of daughter lesions in other fields follows the same trends, it must also be shown that the multiplication of disease in the field of origin is not appreciably affected by the area of this field. This follows from direct observation. Increasing the area of fields reduces the amount of inoculum that escapes from the field of origin. If this affected the multiplication of disease there, one would expect gradients of disease within the field, the border being significantly less diseased than the center. Small border effects have indeed been observed: e.g., Thomas et. al. (1944) found that peach yellow-bud mosaic occurred less in the outside row or two, when disease multiplied within and the orchard was not exposed to infection from without, and Storey and Godwin (1953) found that when cauliflower mosaic multiplied within a field the incidence was somewhat less in the outer rows. But these border effects are small and extend inward for only a few rows or feet. If they had been strong, they would long ago have received general comment, and one may infer that with the exception of very small plots and fields, the area of the field does not greatly affect the rate of multiplication within. The reason is not difficult to find. The vast bulk of inoculum released from a field falls back into the same field. Variations in the small proportion that escapes have a large effect on inter-field movement of disease, but very little on disease within the field of origin itself.

F. The Paradox; in Praise of Large Fields; the Second Rule of Sanitation

The paradox is this. Bringing plants together into fields increases the chance of epidemics; bringing them still further together, by increasing the area of the fields and correspondingly reducing their number, may reduce the chance of a general epidemic.

In the literature of plant pathology it is common to read of the danger of bringing together the host plants of a pathogen. The indiscriminate indictment for bringing plants together is unjustified. Food must be grown, and it is time to write in praise of the large field.

Take these two examples: swollen shoot of cacao can be controlled by sanitation, i.e., by cutting out diseased trees. But the scope for this on small farms is limited, and the difficulties have been described by Posnette (1953). "Each farm, consisting usually of less than 5 acres, has to be treated separately although it has no clearly defined boundaries, and from the standpoint of disease control about 50 acres is the smallest area it is practicable to consider as a unit. For treatment to have any permanent value, the unanimous co-operation (or at least the consent) of many individual farmers must be obtained, and no method of achieving this has been found. Consequently, the Ivory Coast, Nigeria, and the Gold Coast [Ghana] have each in turn been forced to abandon their original plants for disease eradication and have had to adopt the expedient of a 'cordon sanitaire' around the heavily infected districts."

At the other end of the scale we read of the trend gardening has taken in the Everglade region of Florida in recent years. Anyone farming less than a section or two of land—1 or 2 square miles—is considered a little man in the vegetable patch. Celery fields are half a mile long. A celery harvester has been described that weighs 60,000 pounds and carries a crew of nearly 100 persons aboard. Weeds are cleaned our chemically. The writer does not pause to mention such a trifle as disease, but if our inference from Wellman's findings is correct (in Section IV, E, 1) celery mosaic from weeds must cease to exist as a practical problem.

If inoculum comes from other hosts—as cucumber mosaic virus in celery fields comes from *Commelina* and other weeds—making fields larger reduces the percentage of infection in the fields. If inoculum moves from field to field, making fields larger and correspondingly fewer reduces the movement. If sanitation is practiced within the field—as by using clean seed if the pathogen is seed-borne, by rotation of crops if the pathogen persists in the soil, by roguing out diseased plants, and by planting disease-free nursery stock—the size of field does not affect the process directly. But making fields larger and correspondingly fewer affects it indirectly by reducing the reentry of inoculum from without. Guarding against the reentry of large amounts of inoculum is part of the process. One can summarize all this in a second rule of sanitation: efforts at sanitation are furthered by making homogeneous fields, orchards, and plantations larger and correspondingly fewer.

The qualification that the fields should be homogeneous is implicit in all our arguments. If, for example, part of a large field is sown with healthy and part with diseased seed, there is no necessary advantage in largeness; it would have been better to separate the two parts. Sanitation needs common sense as well as common rules.

With diseases that spread slowly, such as some root diseases, the second rule may seem to have no urgency. But these diseases spread in time, as experience shows, and the rule applies in time.

G. The Reduction of Disease by Farm and Country Planning

What is discussed under this heading in five paragraphs, namely, the planned reduction of disease by the proper spacing and grouping of fields, may well have in the future a chapter and then a treatise to itself. This is sanitation, no different fundamentally from the destruction of alternate hosts such as barberry bushes. But we are concerned here not with other hosts, but with the crop itself. Two examples will illustrate the problem. One of the chief reasons for epidemics of virus yellows of beet in England is the indiscriminate intermingling of the sugar and fodder beets with some 8000 acres of seed crops in areas naturally congenial to the aphid vector. Concentrating seed growing into larger fields in fewer regions would reduce the chance of epidemics. In Holland 80% of the surface of the area, De Streek, is down to potatoes, mainly the early-maturing variety Duke of York, which is intensely susceptible to blight. This (presumably) has come about entirely for reasons unconnected with blight, such as the suitability of the soil and climate for early varieties. But the fact remains that the concentration of very susceptible early varieties is excellent planning of a countryside against blight, because the over-all damage is less than when very susceptible early and less susceptible mid-season varieties are intermingled.

The problem is to compute the effect of a change in the spacing and grouping—the pattern—of fields. The simplest way is to put matters on a relative basis, to use the existing pattern as a standard, and to compute how a change in the pattern would change the date of the onset of an epidemic. From the gradients one can determine the percentage change in the number, i.e., the relative number, of daughter lesions that a field acquires from other fields. From the percentage change one can use equation 11 to determine Δt, the delay in the onset of the epidemic, i.e., the relative date. If, to quote for illustration purely random figures, Δt is 6 days, the date of onset might be postponed from July 20 to 26, or from August 1 to 7, or from November 12 to 18, the actual date, as distinct from the delay, will not be estimated in advance. But the estimate of the delay itself is enough for an estimate of the gain in yield (from a knowledge of average dates of onset and average increments of yield with the development of the crop). From this one could judge whether any feasible change in the pattern of fields is worth recommending.

It is not implied that the calculations would be simple, but ways and means of numerical computation can usually be found. The determination of gradients, especially over distances not near the source, may be difficult; among fungi, a start might be made with those that release their spores only in the cool of the night and early morning or during cloudy, wet weather: their gradients are likely to be more easily determined. But collecting an appropriate range of r values for different varieties, different maturities, and different conditions of the weather should not be difficult; for some diseases such as potato blight, data already exist in the literature, from which a range of values could be calculated.

The response of the various diseases will vary. The greatest benefits from a change in the pattern of fields—often a delay of months in the onset—can be expected when the gradient is steep and r is low. (Steep gradients will also simplify the computations because they limit the area that must be scanned). Diseases such as potato blight, with high values of r but steep gradients, are less apt for inclusion in planning but cannot be excluded. Diseases with both shallow gradients and high values of r are probably not worth planning for. This lack of universality does not condemn the method any more than the use of fungicides in plant pathology is condemned by their failure (up until now) to help much in controlling the rust diseases of cereals. One chooses a method by what it will do, not by what it does not do.

The general scope for planning is becoming more favorable under both poles of farming, capitalistic and communistic, because mechanization is introducing larger units are more freedom for organization. On the evidence, there is already room for maneuver, at least with some diseases.

H. Epidemics in Experimental Plots; an Epidemiological View of Field Experiments in Plant Pathology

In Section II, E we discussed secondary epidemics. Potato varieties, which although not immune are resistant enough to escape epidemics on their own, can become severely infected through the multiplication of inoculum received from susceptible varieties. Apples of the variety Delicious have resistance but not immunity to powdery mildew, and may become infected when interplanted with susceptible Jonathan as pollinators. So, too, when one plants small experimental plots of resistant but not immune varieties in the same variety trial as plots of susceptible varieties, one can expect inoculum to move from the susceptible to the resistant varieties and initiate there a secondary epidemic that otherwise would not have developed. If the purpose of the trial is a demonstration of qualitative differences, no harm is done. But if the purpose is to put a quantitative value on the resistance, this value will be underestimated, possibly grossly, and the trial will give no indication of how the variety would behave in the hands of the farmer, which after all is the point of the trial. Smallness of plot magnifies the error because it magnifies movement of inoculum between plots. Also, resistance itself contributes to the error: the greater the resistance, the smaller the value of r, so, by equation 11, the greater the advancement of the date of onset of the epidemic in consequence of the arrival of a given amount of inoculum from the susceptible varieties, and hence the greater the error in evaluating resistance, if the onset is early enough to affect the result. The first rule of sanitation (Section II, C) prevails in experimental plots as much as elsewhere; here it is in reverse, with an advancement instead of a delay of the onset (i.e., with Δt negative).

Precisely the same argument holds in trials with fungicides. If the fungicide were perfect and conveyed complete protection in all circumstances, no harm would be done by the proximity of unsprayed control plots. But fungicides ordinarily fall short of perfection, and the presence of unsprayed controls necessarily causes the value of the fungicide to the farmer to be underestimated. An example has been given by Christ (1957). A fault on the right side, one may say, but it remains a fault.

There has been a strange change of fashion. In the bad old days experimenters dispensed with adequate replication. Their experiments were often biologically sound, but seldom capable of statistical interpretation. Nowadays we determine with great mathematical nicety the statistical significance of a result that has no reality outside the experimenter's plots. It seems that the statistician and the plant pathologist have each wrongly assumed that the other has examined the techniques of field experiments in plant pathology and approved of them. Better techniques could probably be evolved, perhaps by using plots of unequal size and cutting down subdivisions as much as is possible without interfering greatly with statistical efficiency. But the real solution is to reduce the errors at their source. If, for demonstration purposes, it is necessary to plant a resistant and susceptible variety together, or to have a sprayed and unsprayed plot together, it should be considered as a demonstration and left at that. But if one wishes to evaluate new varieties quantitatively, they should be compared with others of about the same class: susceptible varieties with susceptible varieties, moderately resistant varieties with moderately resistant varieties—each class in its own separate trial. In fungicide trials one fungicide should be compared with another without an untreated control, if the purpose of the trial is to determine which fungicide is the more efficient. This will reduce the error, even if it does not entirely exclude it.

From these remarks one excepts diseases like many root diseases that spread too slowly for plots to interfere with each other.

V. THE HOST PLANTS

A. Conditions for an Epidemic of Disease in Annual Crops

1. *The Annual Rise and Decline of Epidemics: Rules about Its Form.* Among annuals we include perennials and biennials grown as annuals, such as beet grown for its roots.

Epidemiologically, the problem with annuals is to determine how disease can strike quickly enough to cause substantial damage. Heavy losses can occur quickly enough if small lesions multiply fast enough to compensate for scarcity of inoculum at the start, or if the amount of inoculum at the start is high enough to compensate for a slow rate of multiplication of small lesions, or if each individual lesion causes damage great enough to compensate for low initial inoculum or slow multiplication, or with any of the countless gradations between these ways.

Consider the first two ways. According to equation 11 the higher the value of r, the less the effect on an annual epidemic of variations in the carry-over of inoculum from the previous season, and hence the less the effect of environmental factors other than those operating during the season of the epidemic itself. The climatic and other risks are concentrated and not well spread, and, as a general rule, one can expect large fluctuations from season to season.

The same rule can be put more elaborately: the higher the value of r, the less likely is a nonrandom succession of epidemic years, provided that the variety of the host and the race of the pathogen remain unchanged. The provision is necessary to exclude what happens when a new aggressive race arises and causes successive epidemics until it is countered by new resistant varieties.

Stem rust of wheat, caused by *Puccinia graminis*, illustrates this rule. The explosiveness of its epidemics indicates a high value of r, and the data compiled by McCallan (1946) show a wide fluctuation—from a trace to 23%—in crop losses from year to year in the United States. Leaf rust, caused by *P. triticina*, which on the whole produces less explosive epidemics, does almost as much damage on the average, but within narrower limits, from 1.0 to 9.6% as one would expect from a lower value of r.

Another rule, which needs little explanation, is: the higher the value of r, the steeper the rise, and, usually, the steeper the decline of an annual epidemic. The second part, about the decline, can only be inferred logically for low-priced crops such as corn and wheat. With high-priced crops that justify a large fungicide account one cannot infer that inoculum drops to a low level between seasons, even if r is high.

2. *The Fast Multiplication of Small Lesions with a Low Amount of Inoculum at the Start.* Because of a high value of r, sanitation is not very effective against diseases with these characteristics (Section II, C) unless it is undertaken thoroughly and extensively, as in a nationwide eradication of barberry bushes for the control of stem rust of wheat. Epidemics are typically controlled by using resistant varieties (e.g., against cereal rusts) or fungicides (e.g., against late blight of tomato). Epidemics are typically widespread; they involve more than a few farms and commonly extend over whole countries or states. This accords with what was said in Section III. Epidemics tend to fluctuate widely from season to season.

3. *The Slow Multiplication of Small Lesions with a High Amount of Inoculum at the Start.* The root knot nematodes (*Meloidogyne* spp.) of tropical and subtropical climates are examples. They survive well from year to year. The multiplication rate is relatively low; e.g., one does not expect severe epidemics if the soil is only lightly infected at the beginning of the season. This is indeed the fundamental assumption in soil fumigation, which reduces the population of nematodes but normally falls far short of complete eradication.

Because of a low value of r, diseases in the category can usually be controlled by sanitation, e.g., by fumigation, isolation, clean fallows, or crop rotation. Epidemics are typically local, i.e., they may involve only a part of a single field. Fluctuation from year to year are relatively small; one can normally predict fairly accurately what will happen one year from what happened the previous year.

4. *Multiplication of Large Destructive Lesions and Systemic Infections.* If each individual lesion or systemic infection is very destructive, relatively slow multiplication of a relatively small amount of inoculum at the start will bring about a destructive epidemic.

Because of slow multiplication, sanitation is commonly effective (Section II, C). Thus, for virus diseases of annuals, sanitation is the commonest recommendation for control. An analysis of recommendations in the 1957 edition of Smith's "Textbook of Plant Virus Diseases" establishes this point. Epidemics tend to be local (Section III, G). Evidence here is strong; there does not seem to be a single example among annual plants of a disease with large lesions that spreads widely in a year without man's help. One expects disease not to be fluctuate wildly from year to year. Evidence here is weak and conflicting. Hull (1952) states that epidemics of virus yellows in sugar beet develop in a spiral over a period of years. Peak outbreaks have never yet occurred in the root crop in England directly after years of light infection, but have taken 2 or 3 years to develop. But Wolf (1935) found no correlation between the amounts of mosaic in consecutive tobacco crops on the same land, even though in 125 out of 229 fields the virus was observed to overwinter in tobacco stubbles.

B. Conditions for an Epidemic in the Annual New Growth of Perennial Crops

There is the same need for speed here; the new growth must be infected quickly enough for the epidemic to be destructive.

Some diseases owe little to the perennial nature of the host; e.g., tuber-borne inoculum does not usually play a significant part in epidemics of early blight in potatoes, caused by *Alternaria solani*. The greatest effect of perenniality occurs when the crop occupies a perennial site such as an apple orchard or vineyard or when the disease affects perennial tissues as well as the annual new growth. Inoculum then commonly overwinters locally; e.g., epidemics of apple scab start from inoculum released in spring from rotting leaves or infected twigs; epidemics of apple and grape powdery mildew start from infected bud scales.

Many of the diseases of the annual new growth of perennial crops are caused by fungi that multiply fast in small lesions. They differ from corresponding diseases of annual crops in often having a larger initial source of overwintered inoculum. Consequently many tend to be particularly destructive, and must be kept under control by the lavish use of fungicides. It is a matter of fact that the greatest sales of fungicides (excluding seed dressings) are for use against diseases in this category. The crops must be productive enough in terms of money to pay the bill.

C. Epidemics of Disease in Perennial Tissues

1. *Plants Grown from Seed.* The diseases in coniferous forests fall within this group, as well as many diseases of hardwoods, plantation crops, ornamentals, and the like.

In agriculture one is concerned largely with epidemic disease; in forestry, largely with endemic disease. With indigenous pathogens of indigenous trees growing under natural conditions, inoculum has existed over a long period, and a balance is struck between pathogen and host. Disease control is primarily a matter of: *a*, forestry practice to maintain a reasonably healthy balance; *b*, selecting suitable sites for species adapted to the locality with stok grown from seed of suitable origin and with the appropriate composition and management of the stand. There may be small local and temporary epidemics, but the general ecological pattern is of a stable community of pathogens and hosts. In the language of tern is of a stable community of pathogens and hosts. In the language of this chapter there is no multiplication and no spread; inoculum is perennial, and within limits the area of the forest has little effect. What we have been discussing is inapplicable—which emphasizes the point that this chapter is concerned only with epidemic disease and not disease in general.

With introduced pathogens or introduced hosts a new balance must be struck, and it is commonly very unfavorable to the host. Examples are chestnut blight, Dutch elm disease, and swollen shoot of cacao. Epidemics of disease following new combinations of host and pathogen seem to be much like those of disease in annual plants, except that the time scale is different; one can conveniently plot the progress of the epidemic in years rather than weeks.

2. *Dangers of Vegetative Propagation. a. Special Danger from Systemic Disease.* Although the material used for vegetative propagation can carry to inoculum of local lesions, it is especially dangerous for systemic disease. In "Plant Diseases", the 1953 Yearbook of Agriculture of the United States Department of Agriculture, about 8% of the space given

to diseases of plants grown from seed is about viruses. For vegetatively propagated plants the figure is 28%. It is largely because of the property of systemic infection that viruses are so dangerous in plants propagated vegetatively.

b. Danger from Longevity. Many of the clones used in agriculture and horticulture are old. The bulk of the citrus fruit of commerce comes from varieties 80 years old or more. Old apple varieties remain highly popular. The potato varieties Russet Burbank, Irish Cobbler, and White Rose, which rank second, fourth and sixth, respectively, in the United States, are all very old. This longevity provides the span needed for a slow epidemic—what in Section I, B was called a low death rate epidemic. It is one of several factors in the accumulation of virus diseases, so that, for example, it is rare to find a single healthy clonal citrus tree, as experience with the Florida budwood certification scheme shows.

The longevity factor (the danger of perpetuating the virus) is the one usually considered in references to the danger of virus diseases in vegetatively propagated plants. But it is easy to overrate the importance of the factor. In providing a long span of years long-lived clones are no different from ling-lived plants grown from seed.

c. Danger from the Randomization of Sources of Infection. Figure 2B shows a nest of leaf-roll-infected potato plants loosely clustered about the plant that was the original source of inoculum. Within the nest, and especially along the rows which are the main direction of spread, there are many nonrandom contacts between diseased plants: contacts that are harmless in the spread of a systemic disease. Within the nest diseased plants are separated from healthy plants outside the nest; the deeper they are within the nest, the greater the separation and the less the chance of transmission. Suppose that the field of potatoes had been harvested and the crop used for seed the next year. There would have been a mixing of seed with a scattering of diseased tubers among healthy ones, so that when the new crop grew there would be new random contacts between diseased and healthy plants that would allow a less hindered spread of infection. The phenomenon is general. When systemic disease spreads from a focus in an orchard to form a nest of diseased trees, there is a similar automatic brake on further spread and a similar release of that brake every time buds are randomly collected as propagating material for a new orchard.

The danger from this source is greatest when r or r' is small, as it usually is with systemic disease of trees in particular. For reasons given in discussing scales of distance in Section III, E the smaller the value of r', the greater the tendency for diseased plants to cluster, hence the stronger the automatic brake and the greater the benefit to the pathogen of a random use of propagating material in establishing a new orchard or field. One could paraphrase this in general terms by saying that the feebler the pathogen's own power of spread, the greater is the relative benefit to it of man's moving propagating material around, and that feeble powers are on the whole likely to be found in systemic pathogens of trees.

d. Danger from the Infectious Incompatibility of Scion and Rootstock. One of the worst epidemics on record, that of tristeza disease of citrus in South America, resulted from what in the event proved to have been an unfortunate choice of rootstock. Typically, tristeza was found in sweet orange on sour orange rootstocks although other combinations were involved. Neither sweet orange on sweet orange stock nor sour orange on sour orange stock suffered noticeably. The disease was one of a combination of species, not of single species.

The key to the understanding of tristeza is the difference between transmission by grafting and by vectors. Infected sweet oranges harbor with apparent tolerance a virus component which is readily transmissible to sour orange seedlings by grafting, and causes them to be stunted. But although this component is freely transmitted by vectors such as *Toxoptera citricidus* from sweet orange to sweet orange, it has not been found in adult sour orange trees even when they grow beside infected sweet oranges in a vector-ridden orchard. Sour orange in the orchard is resistant to systemic infection by vectors of this particular component. But when one grows a sweet orange scion on a sour orange rootstock in a tristeza-infected locality, the sweet orange foliage acquires the component by vector transmission, and the sour orange rootstock acquires it from the scion across the graft union. The result, it seems, is tristeza.

To generalize, the necessary conditions for infectious incompatibility between species neither of which shows marked symptoms on its own roots are probably that both species should be susceptible by graft transmission, that one or both species should be resistant to systemic infection by vectors, and that one but not both species should be tolerant of the pathogen or particular strain of it. Exocortis of sweet orange on trifoliate rootstocks is probably an example in which both species are resistant to systemic infection by vectors (no vector is known). Incompatibility in such a combination is normally the result of using infected material in the nursery.

e. Pernicious Nursery Practices; Contamination During Handling. It would be unrealistic to ignore the part that pernicious nursery practices can play in increasing diseases which on their own increase slowly. Rootstocks have been rebudded after the first buds have failed, sometimes with a different variety. Rose nurserymen have been known to obtain an abundant source of material for rootstocks by cutting stocks off above the bud union and reusing the cuttings after rooting them for a new lot of buds the following season.

Rather different in type because the practices are not in themselves reprehensible is the spread of inoculum while handling the material for propagation. Ring rot, caused by *Corynebacterium sepedonicum*, black leg, caused by *Erwinia phytophthora*, and the virus disease spindle tuber are transmitted by the knife used to cut seed potatoes, and, other things being equal, epidemics are more common when seed is cut and not planted whole.

f. Genetic Uniformity of Clones. Apart from mutations, a clone remains genetically uniform. If a pathogen can attack the clone, genetic conditions are uniformly favorable to disease—a point sometimes stressed. On the other hand, if the clone is resistant, conditions are uniformly unfavorable to disease—a point sometimes overlooked. The problem is part of a wider problem discussed in Section V, E. The danger of secondary epidemics in crops not uniform in resistance has been mentioned in Section II, E.

g. Disease in Relation to Vegetative Propagation in Nature. Plants propagated vegetatively are common in nature, ranging from sod grasses to suckering trees. As a group they do not seem to be especially liable to disease. In them vegetative propagation is primarily a matter of longevity and (presumably) of genetic uniformity. In particular, there is no randomization of sources of infection. Members of the clone stay close together, even when propagation material is released from the air, as the bulbils of *Agave*. To put the matter teleologically (purely for the sake of brevity!), nature does not make the mistake of distributing this material as she distributes seeds and fruits. Only man makes this mistake, and gives the pathogen a mobility it would otherwise not have.

The dangers of vegetative propagation usually stressed in the literature are the dangers of perpetuating the pathogen in the clone, i.e., the longevity factor, and the danger of genetic uniformity. If those were the worst dangers, the prospects of improvement would not hopeful, because the dangers would be inherent in the core of the process of vegetative propagation. But with vegetative propagation in nature as a background one doubts whether they are. An assessment of disease factors in vegetative propagation in agriculture and horticulture seems overdue. In particular, it should be assessed whether vegetative propagation is in fact inherently dangerous or whether we are not making it more dangerous than it need be. If, as one might guess, much of the trouble starts from the randomization of sources of infection, it should not be beyond the ability of man to devise means of curbing the randomness. Partial curbs are already being applied: perhaps the best-known is tuber-uniting of potatoes (cutting seed potatoes and planting the pieces together in sequence). But more could be done, and pathologists might take a broad look at the disease problem in vegetative propagation to see whether they cannot cut it down to size.

D. Epidemics of Systemic Disease

1. *Relation with the Size of the Plant.* In Section III, F a systemically infected plant was regarded as a single large lesion. Between small plants growing close together in narrow rows inoculum has less distance to travel than between large trees proportionately widely spaced. Other things being equal, an individual vector, for example, will transmit disease more easily between the small plants than between the trees. But there in another quite different problem. If large plants are scattered among the small plants without change of spacing, the large plants are a bigger target for the inoculum and are more likely to become infected. Because it is implicit in the concept of systemic disease that only one successful transmission is needed to infect the plant, these large plants will be expected to develop a larger percentage of systemic disease than the small plants among which they are scattered. Some figures of Broadbent (1957) can be quoted. In cauliflower seedbeds the infection will cauliflower mosaic was 37.5% in large plants, 6.5% in medium plants, and 0.5% in small plants. The corresponding figures for cabbage black ring spot were 13.8, 7.0, and 2.4% respectively.

2. *Relation with the Number of Plants per Acre.* Suppose plants are being infected with a systemic disease caused by inoculum coming from outside the plants, e.g., plants infected with air-borne or vector-borne inoculum entering the field from another field or plant infected with inoculum from the soil in which they grow. Consider, for example, two

fields exposed to a uniform invasion of migrating infective insects. If m_1 and m_2 are the mean number of transmissions of inoculum per plant in the two fields and x_1 and x_2 the proportions of disease developed,

$$1 - x_1 = e^{-m1}$$

$$1 - x_2 = e^{-m2}$$

according to the Poisson theorem. Hence,

$$\frac{\log (1 - x_1)}{\log (1 - x_2)} = \frac{m_1}{m_2}$$

The mean number of transmissions per plant can reasonably be assumed to be proportional to the number of vectors per plant, and, because the vector environment of the two fields is uniform, this is inversely proportional to the number of plants n_1 and n_2 per unit area in the two fields. Thus,

$$\frac{m_1}{m_2} = \frac{n_2}{n_1}$$

and

$$\frac{\log (1 - x_1)}{\log (1 - x_2)} = \frac{n_2}{n_1} \tag{15}$$

This relation (in a slightly different outward guise) was used by van der Plank and Anderssen (1945) for the discussion of infection of tobacco with tomato spotted wilt virus. The virus is brought into fields by thrips which apparently settle randomly and do not spread the disease from tobacco to tobacco. The percentage of disease varied with the number of plants per unit area in the manner predicted by equation 15.

If the proportion of infection is low for a given disease and vector, one may simplify matters by an algebraic transformation, and, as an approximation, write

$$\frac{x_1}{x_2} = \frac{n_2}{n_1}$$

If μ_1 and μ_2 are the number of infected plants per unit area in the two fields, then

$$\mu_1 = n_1 x_1$$

$$\mu_2 = n_2 x_2.$$

So, as an approximation when the proportion of infection is low,

$$\mu_1 = \mu_2. \tag{16}$$

With systemic disease, if inoculum enters the crop randomly and if the proportion of infection is low, the number of infected plants per unit area is constant and independent of the density of the stand. This relation was observed experimentally by Linford (1943) without a realization of the condition that the proportion of infection must be low. Pineapple plants were spaced 12, 15, and 18 inches apart in the row, giving populations of 21,780, 18,150 and 14,520 plants per acre, respectively. The same number of plants per acre developed yellow spot, caused probably by tomato spotted wilt virus and carried in randomly by thrips. The percentage of infection was low (3.3, 4.6, and 5.2% for the 12, 15, and 18 inches spacing, respectively), so the condition underlying equation 16 is satisfied, and the number of infected plants per acre was 720, 820, and 760, respectively.

In modern agronomy there is a tend toward high plant populations per acre. From the point of view of disease levels this is a trend in the right direction for crops menaced by vascular wilt diseases and other systemic or quasi-systemic diseases. A simple experiment makes a useful demonstration: tomato seedlings growing sparsely spaced in trays are easily infected at suitable temperatures with bacterial wilt when a culture of *Pseudomonas solanacearum* is poured over the soil, but it is much more difficult to infect a high proportion of tightly crowded seedlings.

PART B
FORECASTING EPIDEMICS

Man found his crops destroyed by rusts and smuts, mildews and blights long before he recognized the microscopic pathogens. Some men correlated disease with sinfulness; others, finding only a slight variation in the amount of sin and a large variation in the weather, obtained a better correlation—that between weather and disease. Consequently, they called the weather the cause of plant disease. In this imperfect state of knowledge they were able to forecast disease outbreaks before they were able to name the pathogenic fungi. Now, in our still imperfect state of knowledge, we too can forecast disease outbreaks from our science of the interaction of host, pathogen, and weather.

The forecasting of epidemics is a contribution to the forecasting of crop yields. Yield forecasts are useful if they arrive sufficiently early to permit adjustments in acreage or transportation or if remedies can be applied to prevent the prophesied disaster. An empirical relation between weather and yields can assist in adjustments. The relation becomes more logical and satisfying and is productive of remedies if it also includes a knowledge of the contributions of soil fertility and of varieties to yield. Of primary interest here, the relation is more logical and fruitful if it includes a knowledge of the interaction of pathogens with weather, soil, and host; this is most fruitful if a remedy can be applied after the forecast is made.

Reviews of the forecasting of epidemics of plant disease have appeared in the past. Because the weather relations of diseases are often obvious, the practice is a venerable one, and the many examples over the world have been examined by Foister (1929) and by Miller and O'Brien (1957). These reviews provide complete coverage. Therefore, the subject will not be neglected if the present examination takes a different path: the etiology of plant disease has been examined for opportunities for prognostication of disease and subsequent loss. Illustrative examples will be cited where possible.

The forecasting of epidemics is quantitative epidemiology applied with courage. The better to make it quantitative, a formal framework is proposed and employed as follows:

The number of spores reaching a host will double as the production is doubled or arrivals are proportional to Q, the number of propagules released. The arrivals also increase as the trapping efficiency p increases; however, the relation is not linear because the disseminating "cloud" of pathogens is depleted by this trapping. Thus, arrivals are proportional to $pe^{-\,constant\,pf(X)}$ where $f(X)$ is a function of distance X. The number of spores per volume of air or other medium decreases rapidly with distance X or arrivals are proportional to $1/X^n$; since dissemination is generally three-dimensional, n is about 2. Finally, the number of infections D is related to arrivals or $D =$ constant $[(Q/K)\,p/X^n]\}\,e^{-\,constant\,pf(X)}$. The K is the proportionality constant between arrivals and infection. For the important case of aerial dissemination, estimates of the parameters have been made: p can be about 1/20, n about 2, the $f(X) = X^{1/8}$.

As the etiology is examined for possibilities for prognosis, these possibilities are evaluated in terms of the above model.

I. THE PRIMARY INOCULUM

Many factors will affect the magnitude of the explosion called an epidemic, but the spark that ignites, i.e., the primary inoculum, must be present. And, as the model shows, the greater the quantity Q, the greater the epidemic likely. Thus, the quantity of primary inoculum can be a criterion for forecasting, a criterion particularly useful because of its early appearance.

The diseases of crops grown in the temperate zone have been the objects of most forecasting schemes. Here the pathogen often must assume a resistant form, take shelter, become dormant, or be eliminated and renewed from the tropics. The forecaster can examine these survival mechanisms for his first hint of the new season's prospects.

An early indication of the amount of overwintering primary inoculum is the severity of disease in the previous season. Although many factors can modify the amount, a persistence in amount is seen from year to year in certain diseases due in part to the greater primary inoculum carried over after an epidemic year. For example, the persistence of the catastrophic severity of potato later blight from year to year increased the disastrous consequences of the appearance of *Phytophthora infestans* in Ireland in the "hungry forties." The appearance of the new *Helminthosporium victoriae* led to the persistent decimation of the varieties derived from Victoria oats and necessitated the introduction of new varieties. Thus, the "grand cycle" of disease, the steady increase and decrease of severity from year to year, caused partially by the overwintering of inoculum, is an early straw in the wind indicating the prospects in a new season.

If an overwintering, resistant—often perfect—stage of the pathogen is known, the procedure is more satisfying than complete dependence upon the "grand cycle." Here the forecaster can search for the overwintering form itself and see the spark that may ignite the coming explosion. Many classic schemes for forecasting apple scab proceed by this means. The pathogen *Venturia inaequalis* overwinters in the perfect stage in fallen leaves where its maturity is dependent upon the winter and early spring weather. Microscopic examination of the perithecia, ripening in the warmth of the laboratory, and considerations of winter weather have formed the bases for prediction of the first discharges of ascospores in the spring. Forecasts of infection then can be made from a knowledge of the stage of tree development, the response of growers to the spray warnings, and subsequent discharge and infection periods.

Brown rot of peaches, caused by *Monilinia fructicola*, is another ascomycetous pathogen that overwinters in the perfect stage although the imperfect stage is also functional. A search of orchards for the cankers and mummified fruit that harbor the pathogen reveals primary inoculum for the succeeding season, and could form the basis of a forecast.

Ergot of cereals, especially rye, is another classic disease that overwinters as an easily visible, perfect stage. The discovery of a multitude of the sclerotia of *Claviceps purpurea* in seed or even in a field can be a warning of an epidemic in the field to be planted or in the field where sclerotia may have fallen.

A pathogen may take shelter during the cold season, and a forecasting scheme can be founded upon the amount sheltered. For example, *Bacterium stewartii*, the pathogen of bacterial wilt of sweet corn, survives the winter in the bodies of adult flea beetles. The survival of the sheltering beetles is encouraged by warm winters. Stevens (1934) found that severe epidemics of wilt followed warm winters, and, hence, founded a forecasting scheme employed successfully by him and Boewe in Illinois and by other workers in New Jersey and New York.

A pathogen can invaded seed and be sheltered for the winter with the crop. *Phytophthora infestans* infects tubers in the autumn and overwinters in the storage bin. Here the inspector can estimate the amount of primary inoculum available for the following season. Wallin (1956) has done this with some success. Of equal importance when blighted tubers are culled from the storage and dumped, the sprouts on the cull piles are a fruitful source of inoculum and an opportunity for the forecaster to take a census of primary inoculum.

The extent of virus infection can be accurately predicted by indexing potato seed, a practice generally followed.

Loose smut of wheat or barley derives from seed infected by *Ustilago tritici* or *Ustilago nuda* during the previous season. Thus the census of the infection in the seed can lead to a forecast of epidemics; the detection of infection can be made by germinating or by examining microscopically a sample of seed. A forecast can also be based upon an even earlier event, the weather at blossoming times in the preceding season, because this is the time of susceptibility.

Some pathogens lie dormant awaiting the arrival of warm weather and a host. Their numbers are an index of epidemics to come. This is the case of many soil-borne pathogens—pathogens that spread relatively slowly. The extent of *Phymatotrichum omnivorum* and its depredations can be predicted almost to the foot from its extent in the previous year. Nematode infestation can be forecast because of the persistence of the worms or cysts. Whether or not a crop will be infected by *Verticillium albo-atrum* has been predicted from the response of an index plant, the tomato, grown in soil samples from the intended field. If rotation decreases the population of a pathogen, severity of infection can be predicted from the intensity of management.

Pathogens can lie dormant in an overwintering crop. If the crop, like winter wheat, is harvested in midsummer, the pathogen must race against time if an epidemic is to occur. Hence, the overwintering fungus and its early multiplication are critical. Thus, Chester (1946) found that *Puccinia rubigo-vera* overwinters in Oklahoma in pustules of wheat leaf rust; the level of infection on April 1 is determined by the weather in the late winter; and this level of infection determines the extent of any epidemic, because weather after that date is rarely limiting. Forecasts by this method have been eminently successful, the only failure being caused by the rare event of later weather limiting infection.

A pathogen may be largely eradicated from a region by the winter, and primary inoculum may be borne in on southerly winds. Following the frequent observation of fungal spores in the upper atmosphere, much attention has been devoted to this concept. For example, the annual renewal of *Puccinia graminis* var. *tritici* and the production of epidemics of stem rust of wheat has been attributed after extensive studies by Stakman and co-workers to spores borne aloft in Mexico and showered down upon the Great Plains. Epidemics of tobacco blue mold in Quebec have been associated with the production of *Peronospora tabacina* in Kentucky and a forecasting scheme devised accordingly. Two considerations should be kept in mild in evaluating the foregoing. First, as spring progresses northward, even diseases of local origin will appear to move northward. Second, the rapid dilution by distance of spore clouds, as evidenced by the steep gradients of infection about isolated sources makes the probability extremely small that spores from continental distances will alight on a given field. The second consideration renders spore traps of little use in forecasting; Rusakov found no spores upon traps until 1% of the plants in the neighborhood had been infected, and he suggested a local field inspection instead of dependence upon traps.

The characteristics of stem rust make it more susceptible than many diseases to a forecast based upon spore movement from the South. The astronomical number of spores produced in the large acreages of wheat in the South and the large acreages of hosts to the North somewhat counteract the dilution by distance. The sturdiness of the *Puccinia* spores assures that they will be viable when they arrive. The broad and nearly continuous belt of wheat extending from Texas to Saskatchewan permits *Puccinia* to leap, frog-like from country to country, never requiring it to move continental distances at a single jump. Thus, a forecasting scheme for stem rust can logically be based upon the systematic movement of primary inoculum from the South. The sensitive spores of the downy mildews, produced on limited acreages and seeking out limited acreages separated by miles deserted of hosts, presents a contrasting picture.

Thus, a survey of the inoculum available at the beginning of a new season provides a rational point of departure for a disease forecast. It has the advantage of earliness, with the consequent opportunities for remedial measures, such as selecting an alternate crop, roguing, or chemical control. It has the disadvantage of occurring so early that many factors can subsequently modify the outcome. Nevertheless, the forecaster courts disaster if he does not issue a word of warning when the inoculum potential is obviously high, or, on the other hand, if he forecasts an epidemic solely from weather data when the pathogen is absent.

II. DISPERSAL OF INOCULUM

When a given amount of primary infection is present, the successful pathogen must next produce units such as cells or spores, and have them detached. Garrett calls these units propagules. The product of the primary infection times the output of propagules per lesion comprises the source strength of Q and the subsequent epidemic is proportional to Q. Hence, another opportunity for forecasting is presented.

Here, in considering the production and detachment of propagules such as spores, wheather is met in all of its effectiveness. Many diseases, especially mildews, have long been associated with damp weather; the late blight forecasts of Martin (1923) and Cook (1949) were based upon average or accumulated rainfall and average temperatures. This is, however, an oversimplification. Crosier (1934), interpreting his own and Melhus's studies of the biology of *Phytophthora infestans*, wrote, "It is not the total rainfall nor the average temperature, but the coexistence of moisture and low temperature (from 10 to 15°C.) for one-half hour or longer, that makes possible the formation of swarmspores, and ... infection follows promptly if the moisture persists." Further, he emphasized, "No swarmspores can be formed, irrespective of the temperature and moisture, unless viable sporangia are present." The present discussion, following the life cycles of pathogens, attempts to recognize these fundamentals.

Crosier, in the above study, found sporulation was most abundant in an atmosphere saturated with water and at a temperature of 18 to 22°; sporangia were abundant within 8 hours. Production was slower at lower temperatures, and sterile at temperatures 3 to 5° higher. Here is the necessary information for a description of an environment suitable for the multiplication of *Phytophthora* and the forecast of late blight epidemics.

Quantitative estimates of sporulation of this important pathogen have been made, permitting a more exact weighting of the forecast according to the production. At 21° sporangia appeared in 6 hours, while 8 hours were required at 18°. If more time was permitted, most isolates produced more sporangia at 18 than at 21°. For example, a typical isolate produced about equal numbers of spores, N, in 6 hours at 21° and 8 hours at 18°; in 12 hours at 18° it hand produced about 20 N while at 21° it had produced only 3 N. The forecaster should beware of environmental races, however: one isolate was able to produce between 1 and 2 N sporangia in 12 hours at the high temperature of 27°. Employing data of this type and a survey of primary infection one can estimate the production of spores from hygrothermograph records, and a forecast can be begun.

The final step in the estimation of Q, the number of propagules dispersed, is an examination of the discharge of the units into the air. The units have been attached to the host for nutrition; now they must break this attachment, pass through the enveloping 100 microns of the laminar flow layer of air, and reach the turbulent air above. Many viruses are borne by insects, and their numbers at critical times in the host's life can form the basis of a forecast of epidemics. Some bacteria are borne in splattering raindrops; the arrival of rain might be employed in forecasting a bacterial epidemic.

Fungi have developed many clever devices for dispersing their spores—devices that have been studied by deBary, Buller, and Ingold. Among these are some that depend upon the weather and, hence, can be limiting and of interest to the forecaster. Tobacco blue mold is a case in point. DeBary observed how the sporangia of some Oomycetes are discharged as the sporangiophores dry and twirl; Pinckard (1942) found this phenomenon specifically in *Peronospora tabacina*. The disease tobacco blue mold does not increase during prolonged rains; spores are found in the air as dew or rain dry, not while the lesions are wet. From these observations the forecaster of blue mold can learn to look beyond rains and dews and consider the critical hours after the leaves have dried and the spores have flown.

This completes the estimation of the number of spores released, the source strength Q to which the subsequent epidemic will be proportional if other things are equal. This early omen depends upon the quantity of primary inoculum and infection and upon conditions favorable for the production and release of the propagules. The omen appears to the forecaster early enough for chemical control to be initiated, probably not early enough for the selection of an alternate crop or for roguing. Many hurdles must yet be taken by the pathogen before an epidemic occurs, but the presence or absence of spore production is a valuable, early clue.

III. THE TRANSFER OF INOCULUM

The propagules, once separated from its parent, can be carried to an infection court on some object, in the soil or in water, or through the ar. This transfer is generally not under the control of the pathogen, can be limiting, and, hence, is an opportunity for forecasting.

Many contaminated objects such as seed, soil, and machinery are borne by man. The speed and volume of commerce, the wide distribution of hosts—particularly economic species—and the prolific nature of pathogens causes the probability of permanent nonintroduction of new pests into a region to approach zero. A knowledge of sanitation and of quarantine measures, as well as the foregoing factors, permits the forecaster to estimate how soon the inevitable introduction and subsequent epidemic will come.

The annual reintroduction of pests can be forecast in the same way. For example, the coming incidence of potato ring rot, caused by *Corynebacterium sepedonicum*, can be anticipated from the liveliness of its transfer. Is seed inspected? Is contaminated equipment sterilized? Is whole seed used? If not, then an epidemic can be safely forecast whenever the inoculum was introduced into the seed-growing regions during the previous season.

Many viruses are transferred in insects, particularly aphids. As the number of aphids is decreased, transfer—as is dispersal—is decreased. This may be due to a northern climate, e.g., that of Maine or Scotland; the time of the year; an exposed site; or sometimes an insecticide. Whatever drastically reduces the population of insect vectors reduces Q, and a lower incidence of infection can be forecast.

In many cases a sharp decrease in infection is observed at increasing distances from the source of aphid-borne viruses. In another case the sharp decrease observed by Waggoner and Kring (1956) became a broad distribution when the susceptible Green Mountain variety and extremely high insect populations were encountered. Nevertheless, under the usual field conditions of moderate populations of insects and some host resistance, the decrease with distance will be sharp, disease being proportional to $1/X^n$ where n is about 2. This is not surprising; Wolfenbarger (1946) has catalogued many insect dispersal patterns and found them similar. From this the forecaster can derive two useful conclusions: disease severity will be slight at considerable distances from sources, especially in a single generation of the pathogen; and numerous widely scattered sources are necessary for an epidemic.

The spread of insect-borne fungi behaves in a manner similar to the spread of the viruses, providing similar bases for forecasting. For example, infection by *Ceratostomella ulmi*, the cause of Dutch elm disease, increases with the population of bark beetles (which can be limited by insecticides) and with propinquity to diseased trees. Therefore, a forecast could be based upon a knowledge of the number of beetles and diseased elms in the neighborhood.

In undisturbed soil the spread of pathogens can take place through the contact or grafting of roots. The spread of *Phymatotrichum omnivorum* in cotton fields and of *Endoconidiophora fagacearum* in oak forests can occur in this fashion. Because the span of root systems is relatively small, the forecaster should be able to draw an orderly prognostic map for these diseases and see it verified.

Plant pathogens can be carried by water: rain, irrigation, or spray. If these are the channels by which a pathogen travels, its distribution can be predicted to be localized. Faulwetter's (1917) classic study of the dissemination of bacteria by splashing rain demonstrated the short range of this type of distribution. Spores suspended in some fungicide or insecticide mixtures can infect plants; unpublished experiments with *Phytophthora infestans* demonstrated that spray blasts do not increase its spread consequentially. Pathogens that spread in water evidently spread slowly.

Air-borne fungi have been classic subjects for studies in epidemiology as well as forecasting. Basing his analysis upon the statistical theory of the turbulent transfer of matter through air, Gregory (1945) has proposed a hypothesis for the transfer of inoculum; with slight modification this is the relation introduced in an early paragraph of this review. Gregory demonstrated the wide applicability of the hypothesis, infection generally being proportional to $1/X^n$, where X is the distance from the source of inoculum and n is about 2. Thus, one of the strongest clues available to the forecaster of these diseases is the nearness of inoculum: at half the distance, the probability of infection is fourfold. A concrete example can be seen in the map of an epidemic of tobacco blue mold: a few yards from the source of inoculum in an infected seedbed more than 100 lesions were present on each plant; at the far end of the 3-acre crop less than 5 lesions were present per plant; on a neighboring farm only 1 plant in 4 was infected. With the location of the source of inoculum in this neighborhood known, the course of the epidemic could be predicted for given weather and control practices.

A significant modification can be made in the average decrease with distance predicted by the inverse square relation. The spread of the pathogen downwind from the source is more rapid. Naturally, downwind refers to the direction of the wind during dissemination. Thus, the highest rate of infection would be forecast to the west of a pathogen that spread during rains.

The distribution of insect-, soil-, and air-borne pathogens all being commonly characterized by sharp decreases with distance from the source, van der Plank's (1949) analysis of the relative danger from field size is generally applicable. Thus, small fields and evenly distributed sources call for a forecast of a greater epidemic than do large and widely separated fields.

Two clues for forecasting have been pre-eminent in the discussion of transfer: the increased danger due to more vectors and that due to greater proximity to a source. These factors would have slight importance if the disease were permitted to run its course, but this it is rarely permitted among annual crops. Rather, the host and the pathogen race to the end of the season, and a winner is declared without a steady or equilibrium state ever becoming established. Thus, the number of vectors and the proximity of inoculum are important bases for the prediction of the damage sustained when the crop is harvested.

IV. THE TRAPPING OF PROPAGULES

As the peripatetic plant pathogens pass over a unit area of the field, a proportion p becomes attached to the soil and to plants. An increase in this attached proportion leads to an increase in the number of infections D. Increased trapping exerts another influence: it depletes the cloud of pathogens at a more rapid rate. Thus, D is proportional to $pe^{-\,constant\,pf\,(X)}$, and a knowledge of the rate of trapping p should assist in forecasting the subsequent distribution of infection.

Unfortunately, little is known about the contribution of this factor. Logical conclusions can be drawn from the above hypothesis to supplement our meager knowledge. Contrast the case of a potato infected with Y virus situated among other potatoes attractive to *Myzuz persicae* with the case of a source situated among plants unattractive to the aphid. In the first case, the danger to the adjacent plants is great, but the probability that this nonpersistent virus will be carried afar is slight; in the second case, the danger to adjacent plants is decreased at the expense of more distant ones.

Now let us consider the spores of fungi—objects carried at the mercy of the winds. They vary in density, volume, and consequent settling velocity in the viscous air; presumably this might alter the proportion attached, but experience has shown the effect is inconsequential. The ability of the spores to attach themselves to objects does, however, increase as spore size increases, leaf width decreases, or velocity increases; the small or slow-moving spores drift around a broad leaf. From these considerations more infection nearby, less at a distance, has been forecast for large-spored pathogens in tall grass; and more infection afar, less nearby, for small-spored pathogens in low, broad-leaved plants. Verification is lacking but the logic is compelling.

More than a forecast of distribution can be assisted by the foregoing. We have seen in preceding sections the necessity of a nearby source of propagules if infection and losses are to be consequential at harvest. From this it follows that widespread sources and subsequent epidemics should be frequent with the small spores pathogenic on broad leaves. This conclusion, too, is worth testing with a view to its eventual use in forecasting.

Rain is a well-known cleanser of the atmosphere, washing spores out of the air and onto foliage and soil. This effectively increases *p*, the trapping efficiency. Rain has thus been an important factor in the consideration of long distance dissemination of spores: spores of, say, *Puccinia graminis* are carried aloft and then northward in a maritime tropical air mass, eventually being washed to earth by precipitation. The forecaster must bear in mind the dilution of the spore cloud with distance, but conceivably synoptic charts of the air currents at several levels could be useful in predicting the transfer of prolific fungi that infect large acreages.

V. INFECTION

The pathogen has avoided many pitfalls before alighting, but it now has another crucial step to accomplish: it must infect a plant. In our hypothesis the number of spores deposited per successful infection is *K*; infection increases as *K* decreases.

The first prerequisite is that the propagules be alive. The forecaster will have investigated the hardiness of the pathogen and entered this into his "rules." For example, he will be confident that the hardy rust spores or apple scab ascospores or the virus safe within an aphid are alive, and that this step provides no opportunity for a forecast criterion. On the other hand, he knows that the spores of the downy mildews are susceptible to drought and that sunlight is a signal for a forecast of "no epidemic."

Germination is the second prerequisite for the success of a spore. This step is frequently susceptible to weather and has been a most fruitful source of forecasting methods. Satisfactory conditions for germination must arrive while the spore is viable. The requirement for moisture is common and has been widely applied in forecasting, e.g., the downy mildews. In contrast to this are the powdery mildews with thier requirement for high humidity but not water presumably, this could be employed in forecasting.

The temperature is also critical during germination. Not only can the temperature itself be limiting, but it determines the speed of germination, and hence, the number of hours of satisfactory moisture conditions required. Thus, varying critical lengths of moist periods can be set for varying temperature.

Next, the landing and germination must occur on a host. This success is improbable if few of the host species grow near the source or if the plants are small and cover little of the soil. In this case few of the spores will alight on a host, and many will fall on strange species or barren soil. Here, *K* is large, since many spores fall and few infect; the forecaster may be conservative in his warnings.

Finally, the host that receives the pathogen must be susceptible. In the case of a heteroecious fungus, epidemics are most likely if the repeating stage is on the economic crop. Stakman (1942) has provided us with examples: *Gymnosporangium juniperi-virginianae* sporidia alone can infect apple, and no epidemic of rust would be forecast far from red cedars; *Puccinia graminis* var. *tritici* uredospores can infect wheat, and the pathogen can spread through a wheat belt by repeated, relatively short jumps with scattered sources appearing near many hosts.

In any case the strain of the pathogen must be capable of infecting the variety of the host. One of the forecaster's most important indicators is a survey of the various races present and the varieties being grown. Thus, the appearance of Race 56 of *Puccinia graminis* var. *tritici* and the larger acreages of Ceres wheat foretold the elimination in the midthirties of that hitherto resistant variety.

The damage wrought by an infection could be the *D* of our hypothesis. If this were our thinking, then the proportionality constant *K* would be the number of propagules deposited per unit value destroyed. Then *K* would fall, and the probability of damage would rise if we were concerned with a systemic disease, particularly of a large host. The effect is increased further if the individual plant is valuable. Consider how the probability of damage rises as turn from the spots of apple scab to the systemic infection of Dutch elm disease or oak wilt of a prized shade tree.

In the preceding sections we have dealt with the several steps in the multiplication of an air-or insect-borne pathogen. The life of a soil-borne fungus may be less complicated, but an infection must be accomplished and the pathogen flourish within the host if an epidemic is to ensue. The separation of infection and incubation is incomplete in our knowledge but largely unimportant for this discussion. Seedling diseases are frequently associated with slow emergence and retarded early growth. Thus, the occurrence of damping-off of corn by *Pythmin* spp. can be forecast for cold, wet springs. The more specific, soil-borne wilt fungi also prosper best in certain environments: *Verticillium* wilt of tomatoes can be forecast for cool and *Fusarium* wilt for warm climates.

Infection phenomena are often susceptible to the weather and provide a wealth of opportunities for forecast criteria. Can the pathogen survive, germinate, and infect in the environment of the host? If the forecaster answers these in the affirmative, he knows that a frequent barrier to epidemics has been removed and that he had best beware.

VI. THE INCUBATION PERIOD

The rapidity of increase of a pathogen has already been mentioned as critical to the development of epidemics and consequential losses. This is so in large part because the pathogen and host rarely reach equilibrium during the life of an annual crop. Rather the pathogen is racing against the time when unfavourable weather will arrive. Therefore, what Chester (1950) has called "tempo" is critical.

The length of the cycles as well as the successful infections per cycle determine the tempo. The discharge, transfer, and trapping of inoculum and the infection of the host can, and often do, occur within a few hours. Incubation is generally a more extended process and considerable variation in its duration is frequent. Here, then, is an opportunity for alterations in the tempo, alterations critical to forecasting, alterations in the rapidity of reappearance of new sources and in their size Q.

With the fungus safely within the moist host tissues during incubation, temperature—not humidity—becomes the important factor. An example of the importance of temperature in the length of the incubation period is provided by *Phytophthora infestans*. Decreasing the maximum and minimum temperatures from 23 and 15°C. to 20 and 10°C. increased the length of the period by one-fourth. This would result in only 5 instead of 6 multiplications of the pathogen per month. Of course, raising the maximum and minimum temperatures to 35 and 25°C. destroyed the pathogen within the host.

A similar example of altered length of period is provided by *Uncinula necator*: the incubation period was halved by raising the temperature from 15 to 26°C.

Peronospora tabacina illustrates the disastrous effect upon the pathogen's prosperity of high temperature during incubation: hot weather causes "dry weather" blue mold lesions which are devoid of spores, and hence, ends the multiplication of the pathogen, Unquestionably many examples of accelerated multiplication by warm temperatures and truncated epidemics due to hot temperatures can be found among bacterial and viral as well as fungal incubations. If the forecaster is aware of these relations for his charges, he can estimate whether the tempo of multiplication is sufficiently rapid for an epidemic to develop within the season.

A second use for a knowledge of the length of incubation is in detecting synchrony of the cycles of weather and pathogen. Imagine that a period of weather unfavorable for sporulation and infection is followed by a day when infection does occur. If a second day of weather suitable for multiplication occurs before the end of the incubation, new infections will be add. If, instead, the second suitable day occurs at the end of the incubation, infections will be multiplied, no just added. Thus, favorable conditions alone are not enough; they must be meshed with the cycle of the pathogen, a cycle timed in large part by the temperature during incubation.

One large class of pathogens need not race with the host to a deadline at the end of the season. These are the pathogens of perennial plants, such as trees. Here ample time is allowed for an equilibrium to be established, and eradication and exclusion are relatively harmless to the pathogen over the long pull. The forecast will, therefore, be in terms of "when", not "if." The forecaster must guess the equilibrium rate from the best knowledge of the susceptibility of the host, the multiplication of both host and pathogen, and any deterrents to spread applied over the entire region concerned. As an example, among the American elms in the climate of Connecticut one can forecast with some confidence that in the neighborhood of 20% of the elms will become infected annually with Dutch elm disease if an insecticide is not applied, 3%, if it is.

The incubation period is lengthy, relative to the other periods of the pathogen's multiplication, and hence provides a wealth of opportunities for the forecaster, who must estimate how the race between an annual crop and its pest will end or at what level the equilibrium between a perennial host and its pest will be reached.

VII. INTEGRATIONS

The large number of possible prediction criteria enumerated above are not only opportunities for improving forecasts, but are also opportunities for hopeless confusion because of their number. How can such complexity be resolved?

Fortunately a study of the biology of a particular pathogen will generally reveal a critical step to which attention can be directed. Chester's (1946) description of the forecasting of wheat leaf rust provides an excellent example. In Oklahoma, March is the critical month. If the weather during this period permits the first cycles of rust renewal, there is enough time during the ensuing season for the pathogen to reach epidemic proportions. After this month, the weather is rarely limiting. Therefore, the weather and rust behavior during this month can form the basis of a forecast of rust intensity 2 to 3 months later.

The well-known Dutch and Irish rules for forecasting potato late blight also are a simplification based upon the existence of a critical stage in the pathogen's life. In the maritime climates of northeaster Europe the weather during incubation is rarely limiting, and adequate primary inoculum is apparently common. Therefore, the multiplication of the pathogen is limited only by the conditions for dispersal, transfer, and infection. Satisfactory conditions for these three steps are defined by simple rules, and hence, late blight forecasting is straightforward in Europe.

In the central United States, drought or hot weather frequently persists long enough to destroy *Phytophthora infestans* in the host. Therefore, Wallin (1958) deletes from his forecasts of blight the effects of earlier favorable periods if unfavorable weather persists for 21 days or more.

The length of the incubation period can be the critical factor in epidemics. Shatsky's modification of Müller's incubation calendar for vine mildew illustrates this point. The incubation period is related quantitatively to the mean temperature, and fungicide spraying is indicated at the end of the periods thus estimated.

An integration with a strong family resemblance to the above devices has been employed in Connecticut for several years for the prediction of tobacco blue mold epidemics. The biology of the parasite has been reviewed by Stover and Koch (1951), and the spores are known to be air-borne in the morning. Bearing this information in mind, the forecaster examines the hygrothemograph, sky cover, and rainfall records, together with a census of primary inoculum from the tobacco region and decides whether the pathogen could or could not have completed dispersal, transfer, and infection. If an infection period is thought to have occurred a "2" is entered on the calendar. If two more periods occur before incubation of the first is complete, 2's are added each time and the numbers entered on the calendar are "4" and "6." If instead the second infection period occurs near the end of the incubation period of the first infection, the number is doubled and a "4" is entered on the calendar. When a drought or heat wave intervenes and sterile lesions appear, the number on the calendar is reduced to "1", and the multiplication process must begin anew. Thus, the forecasting scheme is an estimate of the population of *Peronospora tabacina* derived from a knowledge of its biology, its primary population in seedbeds, and the weather. We have found in Connecticut that damage of consequence will appear in the field when the estimate on our calendar reaches "64." The arrival of this level can be anticipated from the tempo of the disease and the weather outlook.

The preceding examples illustrate how the complexities of the pathogen's biology can be digested into rational forecasting criteria. Therefore, the prospective forecaster need not become lost in a maze on the one hand or rely upon empirical criteria upon the other hand.

VIII. THE USEFULNESS OF FORECASTING

When one speculates about the value of weather, yield, or disease forecasts, no difficulty is encountered in convincing oneself that forecasts are useful, even necessary. The desired to foretell the future is strong. Nevertheless an enumeration of the characteristics that determine the practical usefulness of predictions is healthy, especially if it is made before a forecasting system is initiated.

The first requirement of a forecast is that it be correct. Perfection is rarely possible in forecasting, and the natural tendency is to be conservation, to "overforecast." In disease forecasting this leads to the warning "epidemic" whenever a chance of such exists. At the same time that the forecaster is being "conservative", the growers' other advisors are naturally enough behaving in the same manner: they urge the grower to exercise control measures at all times. Therefore, the usefulness of the forecaster's warning of "epidemic" may be limited because his is but one more voice added to the chorus of voices urging him to repent and spray.

On the opposite side, the forecaster may be more helpful. He may lead the grower to omit costly control measures by forecasting "no epidemic." Before he can persuade the grower to make such a savings he must have established the accuracy of his predictions in the minds of the grower and his advisors and have overcome the natural tendency to be conservative. In France this happy state has apparently arrived: the warning service for vine mildew has led to the omission of two sprays in most years at an annual savings of two and one-half billion francs. Thus, if the forecaster can bring about the elimination of unnecessary and costly measures through accurate forecasting, he can be truly helpful.

The distance that the forecaster can see into the future, as well as his accuracy, affects the usefulness of his methods. If the prediction arrives in time for remedial action to be taken, it is helpful. If it arrives too late for action, it only lengthens the time over which the grower must bear his sorrow. The predictions of epidemics of seed-borne smuts and virus from the conditions during the flowering of the wheat or from an index of potato tubers has already been mentioned. Here the prognostication can be made for many months in advance, and the grower has ample time to remedy the matter.

The predictions of wheat leaf rust and of bacterial wilt of sweet corn extend for several months into the future. Hence, the farmer has time in which to benefit from the forecast by planting an alternate crop or variety. The French warnings concerning vine mildew are of much shorter period; nevertheless, they arrive in sufficient time to be helpful because fungicidal control can be applied or omitted on short notice.

Long-range or climatological predictions of plant diseases have been made informally. Men have recognized that certain climates are conducive to a disease, others are not. Seed production has been located in regions where the climate makes important seed-borne disease unlikely. In addition to this a formal study of long-range or climatological prediction would be interesting. The epidemics of certain diseases can be traced to critical weather events. The probability of these events can frequently be estimated from the series of weather observations that have now reached great lengths in many localities. The knowledge of biology and weather could be combined to produce estimates of the probabilities of disease at various localities and seasons. These "forecasts" should prove useful because of their long, almost infinite, extension into the future.

The discussion of the period of the forecast has brought out the importance of the control method. If the forecast is to provide more than intellectual satisfaction, there must be a control method that can be exercised and the impending epidemic forestalled. The usefulness of the above prediction of smut and virus epidemics depends upon the existence of alternate, clean seed. The usefulness of the predictions of wheat leaf rust and bacterial wilt of corn depends upon the existence of profitable alternate crops or resistant varieties. The benefit from the warning systems for downy mildews is dependent upon the efficacy of fungicides.

Two diseases stand in contrast to the ones we have just examined. Apple scab infection can be eradicated by suitable chemicals. Hence, a census of infection can be used and the need for a forecast is consequently less. Wheat stem rust control at the present depends practically upon the choice of a resistant variety. Consequently, the forecast of a stem rust epidemic issued in the middle of the growing season is nearly useless to the individual farmer because there is as yet no remedy he can apply to forestall the impending calamity.

The use of a forecast necessitates changes in operations. Flexibility is a prerequisite to the employment of prediction. At least two things affect this flexibility. The first is the quantity of labor and equipment available for control measures relative to the acreage that must be treated. If the entire area can be treated in a day or two, treatment can be delayed safely until a forecast is issued. If the entire area requires many days for treatment, however, the machines must be kept in operation continuously; otherwise, a large proportion of the area might be damaged before treatment in response to a warning could reach it. The second thing that affects flexibility is the complexity of the farming organization. If the man who applies the treatment is also the man who decides when to treat and if supplies are readily available, a treatment can quickly follow warning. Alternatively, if a large organization is involved, planning must precede action. Here the confusion that would follow a change in plans following a warning might well cost more than the application of an occasional unnecessary treatment. Consequently, a short-range forecast is not highly useful; perhaps climatological "forecasts" would prove to be.

An important—perhaps the most critical—factor in the usefulness and in the adoption of forecasting is the ratio of benefit to cost. How great is the benefit from disease control relative to the cost of the control measure? If this ratio is large, the rational grower will apply controls whether the disease is forecast or not. If this ratio is small, no amount of urging will induce the rational grower to act.

In the northeastern United States apple and potato growers know that the yield, appearance, and keeping quality of their produce are profoundly affected by apple scab and potato late blight. They also know that including an effective fungicide among the pest sprays they are already applying is inexpensive. Thus, the ratio of benefit to cost is high, and the growers tend to apply fungicides routinely with little regard for the likelihood of an epidemic.

At the opposite pole stands the owner of a large wood lot populated in part by elms. The death of the elms would not be disastrous to the production of the lot. The cost of an annual application of an insecticide to kill the vectors of the pathogen of Dutch elm disease would be large relative to the annual production of the lot. Therefore, the ratio of benefit to cost is small, and the owner is unlikely to apply a control measure, even when he is assured of the imminence of an epidemic of Dutch elm disease.

Between these two extremes lies a region where the benefit: cost ratio is favorable to the employment of forecasts. The example of the vine mildew warning system in France cited above lies in this region of a favorable ratio. The losses from mildew are disastrous. On the other hand, considerable savings can be made by eliminating unnecessary sprays. Hence, the warning system is employed by the growers.

The benefit: cost ratio also depends upon the annual variation in the amount of disease. If the amount is nearly constant from year to year, experience has taught the farmer how to make his own forecast: next year will be just like this year and will have the same benefit: cost ratio and demand the same measures. This leaves little for the forecaster to do that is useful. When the amount of disease is variable from year to year, the forecaster can be useful, for he will predict when the benefit of treatment will be great or small relative to the cost of treatment.

The forecasting of disease is a contribution to the prediction of yields. It logically proceeds from a knowledge of the interaction of host, pathogen, and environment. In this review the etiology of disease has been examined for opportunities for prognostication of disease and subsequent loss.

The magnitude of the primary inoculum affects the subsequent epidemic. As a clue to the size of the epidemic it has the advantage of earliness, the disadvantage of being subject to many influences before harvest time.

The production of inoculum and its dispersal into the medium about the plant is one of the processes upon which weather can exert its influence and is, therefore, a rich source of forecast criteria.

The transfer of inoculum by air and soil, and as a contaminant is for the pathogen a risky business, because it is dependent upon fortune. The forecaster profits from a knowledge of the proximity of the source of inoculum and of the number of vectors during this random process.

Propagules are trapped and their numbers depleted more rapidly by some types of foliage and weather than by others. This is probably important in predicting the distribution of infection about a source of inoculum.

Infection is a late step in the pathogen's multiplication, but its requirements are sufficiently critical to provide a wealth of criteria for prediction. The general requirement for liquid water at this stage provides the basis for several classic forecasting schemes.

The incubation period is generally much longer than the preceding steps, and therefore is subject to changes of length. Thus, it can influence the rapidity of development of the disease and the level attained before unfavorable weather or harvest ends the annual race between host and pathogen.

The myriad influences can frequently be digested into simple forecasting rules because of the limitation placed upon epidemics by a few critical stages.

The usefulness of the prediction of epidemics depends upon the accuracy and range of the forecast, upon the existence of remedies and the ease with which they can be applied, and upon the benefits from disease control relative to the cost of remedies.

7

ANTAGONISM OF PATHOGENS

Various biological factors can influence the living pathogens of cultivated plants and their hosts. The living worlds is a complex system where organisms fight for their existence, often to the prejudice of other organisms. Sometimes, however, beneficent associations are realized. In the natural biological medium, pathogens are submitted to the influence of neighboring organisms during their parasitic life, saprophytic phases, or during rest periods. Even the susceptibility of the host plant may be modified by biological factors.

Sometimes the pathogen may be parasitized by another fungus or bacterium. The latter lie at the expense of their hosts by drawing from them the materials necessary to their nutrition. These are hyperparasites or parasites of the second degree, and the phenomenon itself is hyperparasitism. In other cases, antagonism exists between microorganisms living side by side, some inhibiting the growth of others. This phenomenon is frequent, particularly in the soil. Organisms may compete with one another on the host surface. In other cases, organisms associate together and cause complex diseases. Thus, viruses may infect plants simultaneously, or a virus and another pathogen on the same plant may affect one another. Nematodes are known to act similarly with other pathogens.

Certain animals, generally predators, sometimes act as antagonists by feeding on plant pathogens.

In this chapter the different types of interrelations of organisms will be discussed in relation to their role in epidemiology and control.

I. HYPERPARASITISM: PARASITES ON FUNGI

Certain fungi and bacteria are parasitic on plant pathogens. The genus *Ciccinobolus* includes species of fungi which attack fungi of the family Erysiphaceae. Thus, *C. cesatii* de Barry develops as thin mycelial threads inside the hyphae and conidiophores of *Erysiphe graminis*. It fructifies in pycnidia having unicellular spores. Other species, *C. ewonymi japonici* on *Oidium ewonymi japonici* and *C. asteris* on *Oidium astericolum*, act in a similar way. These fungi can hinder the growth of the pathogen so that its mycelium contracts and shrinks, while the formation of oidia is very much impeded. However, *Ciccinobolus* seldom reduces attacks of the powdery mildews in nature because it seldom occurs in large quantities. *Ciccinobolus* can be cultivated on artificial media, however, and some encouraging results have been obtained by artificial inoculation.

Species of the genus *Darluca* live in the sori of Uredinales. The best-known species is *D. filum*, which is found in the uredospores or teleutospores of many rusts. *Darluca* also is cultivatable in artificial media. This hyperparasite also seems to play but a limited role in reducing epidemics of rust on plants.

The genus *Tuberculina* also includes species which attack Uredinales. *T. persicina establishes* itself in the aecidia of *Puccinia poarum, Gymnosporangium sabinae, G. juniperum*, and in the teleutospore sori of *Endophyllum sempervivi* and *Puccinia vincae*. In these organs it forms violaceous sporodochia. *T. maxima* is known as a parasite of *Cronartium ribicola*. It vigorously attacks pycniospores and thus can reduce or inhibit the formation of aecidiospores. The local and natural establishing of *T. maxima* has been observed in pine plantings attacked by rust, but its extension has always been limited to small area. In culture it develops slowly and sporulates sparsely. Artificial inoculations are not easy to realise.

Among the hyperparasites of Uredinales let us also mention *Cladosporium aecidiicola*, which grows in aecidia—especially in *Puccinia conspicua*—and *verticillium hemileiae*, in which the thin mycelium lives in uredospores of *Hemileia vastatrix*.

Bacteria can also parasitize fructifications of Uredinales. According to Borders (1938) *Erwinia urediniolytica* attacks pedicels of spores. Pon et. al. (1954) studied the parasitism of *Xantomonas uredovorus* on *Puccinia graminis*, which grows on or in uredospores at a high temperature (30°C.) and in a saturated atmosphere. The bacteria, which inhabit the soil, are spattered by rain up to the rust sori on the low part of the plant. The temperature requirements of the parasite must, however, limit its biological role.

Among fungi the genus *Trichothecium* includes species that attack plant as secondary parasites, are antagonistic to other microorganisms, or grow at the expense of plant pathogens. The fruiting bodies of the grape-downy-mildew fungus *Plasmopara viticola* sometimes exhibit a brick-red coloring, due to the growth of a fungus, *Trichothecium plasmoparae*. Its thin mycelium lies beside the conidiophores and conidia of *Plasmopara*, absorbing the protoplasm, and the conidial fluff then becomes matted. This *Trichothecium* is unable to penetrate the host membranes, and so cannot reach the mycelium of *Plasmopara* in the tissues of the leaf. It is favored by hot, wet weather, and fears acidity.

Trichothecium roseum sometimes grows on the stromas of certain pathogens, inhibiting the formation of perithecia and sometimes limiting the growth of the fungus it attacks. Thus, Koch (1934) observed it in summer on the stromas of *Dibotryon morbosum*, the agent of black knot of *Prunus* and succeeded in inoculating *Dibotryon* with it artificially. In autumn *T. roseum* can attack the stromas of *Polystigma rubrum* on plum trees sterilizing the stromas by killing ascogenous elements of the *Polystigma*. By artificial inoculations an important increase of diseased stromas has been obtained. On artificial inoculation of *Polystigma*, *Trichothecium* increased the number of stromas formed, and these later became diseased. This fungus has a similar effect on the perithecia of *Gnomonia veneta*, and Viennot-Bourgin (1949) has found it on the stroma of *Nectria cinnabarina*.

That certain fungal plant pathogens are parasitized in the soil by fungi and bacteria has been demonstrated in the laboratory by growing pairs of microorganisms side by side. Weindling (1932) has demonstrated the parasitism of *Trichoderma lignorum* on *Rhizoctonia solani*, *Phytophthora parasitica*, *Phythium* sp., *Rhizopus* sp., and *Sclerotium rolfsii*. The fungus acts to inhibit or kill the host hyphae by direct parasitism or, alternatively, by antagonism, as will be discussed later. Thus the hyphae of the *Trichoderma* may wind around the aerial hyphae of the host fungus, penetrating and growing inside them. Sometimes, however, *Trichoderma* acts at a distance on the hyphae of another fungus nearby.

The use of *T. lignorum* has been considered to fight damping-off of seedlings during their short period of susceptibility. This method has not been too successful in practice because (*a*) *Trichoderma* is not usually dominant at the soil surface, (*b*) it has a high moisture requirement, and (*c*) it autolyzes readily.

A species of *Papulospora*, isolated from the soil, has shown a parasitism on *Rhizoctonia solani* in artificial medium similar to that of *T. lignorum*. Its hypae wind around those of the host or enter and grow inside.

Rhizoctonia solani, the well-known pathogen of many higher plants, can also attack other fungi, nearly all belonging to the Phycomycetes. Butler (1957) has studied this parasitism on agar media by sowing the host fungus and *R. solani* on opposite sides of a petri dish. When the two colonies meet, the *Rhizoctonia* is stimulated at first, forming a thick mycelial zone at the juncture of the two colonies. This mycelium first invades the host about 24 hours after the two colonies meet, and after 72 hours invasion becomes general.

The parasite attacks the aerial hyphae and sporophores of its host in two ways. It may penetrate into the hyphae of the host and grow rapidly inside or it may wind around and constrict a part of the host. The protoplasm of the encircled hyphae either becomes coagulated or else it becomes less optically dense and vanishes. When this happens the aerial hyphae of the host fungus sink down.

Not all cultures of the host fungus are destroyed, however. It may resist infection by lysis of parasitized hyphae or by isolation of infected hyphae with a partition. The degree of parasitism varies also with the culture media, temperature and light.

Because *Rhizoctonia solani* is itself a pathogen, it cannot be used to fight other pathogens. But these observations suggest how competition among fungi can occur in nature.

Certain species of the genus *Phythium* parasitize adjoining species. Drechsler (1943) showed that *P. oligandrum* could attack *P. ultimum, P. debaryanum*, or *P. irregulare* by encircling their hyphae or by penetrating them. Conidia and young oogonia are also parasitized. The destruction is so rapid that the oospores generally do not get formed.

Fungi can parasitize the oospores of root-rotting Oomycetes. For instance, the imperfect fungi *Trinacrium subtile, Dactylella spermatophage*, and *Trichothecium arrhenopum* destroy the oospores of *Pythium, Phytophthora*, or *Aphanomyces*. Infection is accomplished by successive perforations of the cell walls of the oogonium and oospore, followed by growth within the oospore of branches, sometimes lobed, and of big suctorial organs. The chytrid *Phlyctochytrium* can attack and inhibit the germination of the oospores of *Peronospora tabacina.*

These fungi seem to play an important role in the soil, reducing the germination of oospores and their infection of higher plants. Even in soils where quanties of oospores are found, the percentage of diseased plants after the first infection remains slight, owing to the presence of these hyperparasites.

The sclerotia of certain fungi are also frequently attacked and killed in the soil. *Coniothyrium minitans* has been observed on sclerotia of *Sclerotinia trifoliorum.* By artificial inoculation in certain types of soil 85 to 99% of the sclerotia were killed in 11 weeks.

In other cases it is difficult to determine whether hyperparasitism or antagonism is involved. Thus, Clark and Mitchell (1942) showed that the sclerotia of *Phymatotrichum omnivorum* survived in sterilized soils whether they received organic amendments or not, but in unsterilized, amended soils microbials activity has led to destruction of sclerotia. The action is temperature dependent. Thus, the percentage of sclerotia losing their viability 12, 30, 70, and 90, respectively, for temperatures of 2°, 12°, 28°, and 35° C. Moisture was required.

II. BACTERIOPHAGE

In 1917 d'Herelle presented evidence for a transmissible lytic principle that acted on the Shiga bacillus, showing that the bacteria have their infectious diseases, too. These infectious agents are now known as bacteriophages and are usually considered as ultraviruses. Although pathogens of man and animals have been particularly studied, some work has been done on phages of plant pathogens. Phages have been isolated from fragments of diseased plants, contaminated soils, and from bacterial cultures.

By now phages have been discovered for many phytopathogenic bacteria: *Agrobacterium tumefaciens, Aplanobacter stewarti, Erwinia aroideae, E. carotovora, E. atroseptica, Pseudomonas angulata atrofaciens, P. coronofaciens, P. glycinea, P. lacrymans, P. phaseolicola, P. pisi, P. syringae, P. tabaci, P. xanthoclora, Xanthomonas citri, X. malova cearum, X. orizae, X. phaseoli, X. pruni, X. solanacearum*, and *X. translucens.* In 1953 Newbond found a phage of *Streptomyces scabies.*

To act, a phage must encounter a bacterium able to fix it. Once fixed, the phage enters the bacterial cell and multiplies within it. Eventually, the bacterium is lysed, and sets the phage particles free. They are then able to infest other bacteria.

Unhappily, this lysis is rarely complete in a culture. The appearance of secondary cultures of the host bacterium is frequently noted in the same plate. The bacteria of these colonies usually have the same morphologic and physiologic characters as their parents and generally acquire their resistance to the phage through mutation.

In other cases the appearance of lysogenic strains can be noted, as Okabe and Goto (1954) showed with *Pseudomonas solanacearum*. Lysogenic bacteria are genetically able to tolerate phages that multiply in them under certain conditions. In these bacteria the virus is in the form of a prophage that is neither infectious nor pathogenic. Lysogenic bacteria are insensitive to the phages they produce, thus showing a real phenomenon of permunition.

Phages differ in their action. Some are polyvalent, while others have high specificity. Sometimes two forms of the same bacterial strain differ in their reactions to the same phage. Thus, form *S* of *Pseudomonas phaseolicola* is more sensitive to the phage than form *R*.

In human pathology some success has attended efforts to use bacteriophages in fighting certain infectious diseases. In plant pathology, however, attempts at phagotherapy have been unsuccessful. The first work was done in 1925, when Coons and Kotila (1925) prevented the rotting of carrot slices by *Erwinia carotovora* through adding the homologous bacteriophage. But the results were not clear-cut with *Erwinia atroseptica* on potato slices. Israilsky (1927) thought he had shown that the previous inoculation of a phage of *Agrobacterium tumefaciens* into a plant impeded the subsequent growth of tumors, but Kaufman (1929) could not confirm his results.

R. C. Thomas (1940) treated corn seeds with a phage of *Aplanobacter stewarti*, and then inoculated them with the pathogen. He obtained 18% disease in control plants but only 1.4% disease in treated ones. A bacteriophage, isolated from tobacco leaves invaded by the wildfire bacterium, reduced infection to 50% when dusted on test plant (Novikowa, 1940). Also, Kawamura (1940) reduced infection of tomato by *Pseudomonas solanacearum* through applying bacteriophage to the soil.

Fulton (1950) studied two bacteriophages active against certain strains of *Pseudomonas tabaci* and *P. angulata*, isolating them from diseased tobacco leaves. He compared their lysogenic and other properties, especially their inactivation by cations. As Novikowa (1940) has shown, he obtained a reduction in diseased tobacco seedlings by treating them with a suspension of phage before inoculation. This reduction was greater when he dusted with a mixture of the two phages rather than with one alone.

From infested fields Fulton (1950) obtained more bacteriophages of *Pseudomonas tabaci* and *P. angulata* as the disease levels increased. On this basis Fulton suggests that the phages could limit the amount of disease. This is related to the action of the rhizobiophages, which are in part responsible for the "tiredness" of alfalfa fields, because of their action on *Rhizobium* in the nodules of leguminous plants. Possibly, the persistence of the wildfire bacterium in soils is controlled by bacteriophages. This hypothesis wants a precise experiment.

III. PREDATORS AND ANTAGONISTIC ANIMALS OF PATHOGENS

A good many insects live at the expense of the higher fungi. They grow in the carpophores, eating either the flesh of the fungus or its spores or both. Many species of Coleoptera belonging more particularly to families of Scaphidiidae, Liodidae, Staphylinidae, Erotylidae, Cryptophagidae, are mycophagous. A few species specialize on one type of fungus. For instance, *Gyrophaena strictula* limits itself to the Polyporacae.

A succession of several species of Coleoptera feeds upon one fungus, one feeding at one growth stage, another later on, until the rotting fungus is fed upon by the last in the succession. The coleopterous mycetophiles require moisture. When fungi are normally moist, the number of visitors can be very high, but the number decreases with dryness.

Fewer Hymenoptera attack carpophora. Possibly the Lepidoptera in the imago stage suck the droplets secreted by fungi. Some larvae that cat fungi are known.

The fungus is merely a shelter for the majority of the Hemiptera found on Polypores. The Aradides and Dysodiides are considered as mycophagous and are often seen between the lobes of Polypores.

The inoculum of plant pathogens (smuts, rusts, mildews) can be appreciably reduced by Coleoptera and Diptera that subsist on these fungi, among them being (Phalacridae, Coecinellidae, etc.……) (Itonididae-(Cecidomyidae) Fungivoridae etc……).

Among the representative mycophages of the family of Phalacridae is *Phalacrus caricis* Sturm, observed Ditguilar (1944) in a *Cintractia* head of *Carex riparia.* The larva subsists exclusively on spores of *Cintractia subinclusa,* these spores being evident inside the alimentary canal of the insect. When adults appear at the end of July, they have the same diet of fungus.

Many species of the race *Phalacrus* destroy the smut spores produced or grasses. The spores of these insects have lost in the excrements their viability. Hence insects are not dispersal agents of the Ustilaginaceae. When Friendrichs (1908) placed material drawn out of the posterior intestine on nutrient agar, the contained spores did not terminate. The same was true of spores drawn out of the middle intestine. However, the spores from grain that sheltered larva of *Phalacrus* germinated readily.

The larva and adults consume quantities of spores. This regimen and the abundance of *Phyalacrus fimetarious* on the ears of cultivated plants parasitized by Ustilaginae give to the genus *Phalacrus* a certain economic importance.

Other examples of insect predators of fungi include *Deuterosminthus bicinctus* var. *repanda*, which browses on conidiophores and conidia of the downy mildew fungus of grape, *Plasmopara viticola*.

Finally, let us report the antagonistic action of certain inferior representative of the animals kingdom. In artificial media the mycelium of *Verticillium dahliae* does not grow, and its pseudosclerotia do not germinate in presence of infusoria of the race *Culpoda.* Tomato plants grown in nutrient solution containing pseudospores of *V. dahliae* wither rapidly, although no infection occurs when the solution contains active infusoria. In infested soils when *Culpoda* is present, the intensity of the disease is decreased. The mode of action of the protozoan is unknown.

IV MICROBIAL ANTAGONISMS

A. Phenomena of Antagonism

The phenomena of antagonism between microorganism have been particularly studied in recent years with the development of antibiotic substances. When a nutrient medium is sown with two microorganims, one frequently inhibits the growth of the other.

Different classic techniques demonstrate this phenomenon. They consist of sowing two or more organisms at different points on the surface of an agar medium and allowing them to grow or in inoculating a plate with fungus spores or bacteria and then placing a fragment of agar bearing another organism on the surface nearby. Inhibition zones are then often observed. In the periphery of the zone, morphogenic effects exist. Deformations or the lysis of vegetative elements of the inhibited organism are obsèrved. A short distance beyond stimulation occurs in the form of much-branched hyphae, increased sporulation, etc. Temperature, composition of nutrient media, and other factors affect antagonistic phenomena.

Many microorganism, bacteria, actinomycetes, fungi isolated from the soil, or various substrata can thus act on phytopathogenic agents.

In a study on several thousand isolates 256 of them, including 60 strain or species of bacteria, 90 strains or species of actinomycetes, and 106 strains or species of fungi, have revealed antagonistic action on phytopathogenic agents. The latter included 10 bacterial species and 43 fungal species belonging to various families: Phythiaceae, Ustilaginaceae, Telephoraceae, Erysiphaceae, Heliotiaceae, Sphaeriaceae, Valsaceae, Phomaceae, Melianonaceae, Moniliaceae, Dematiaceae, and tuberculariaceae. Other studies have increased this list, and it is probale that most or the whole of the phytopathogenic agents are susceptible to antagonistic organisms.

In certain cases no distinct zone is established, but the growth of the pathogen is inhibited by the invasion of the hyphae of the antagonist. That may be accompanied by the winding of the antagonistic hyphae around those of the inhibited fungus and sometimes by active hyperparasitism.

Moreover, the action can be mutual. Thus, *Alternaria solani* inhibits the growth of *Streptomyces griseus*, but its growth is also inhibited and its mycelium deformed by the influence of *Streptomyces*.

The liquid nutrient medium on which an antagonistic organism has grown can also inhibit the growth of a pathogen. This can be demonstrated by the disk or link method.

Waksman (1957) has discussed how antagonism arises. In the competition for space the vegetative vigor of one organism can impede the growth of another (many Mucoraceae).

Also, physicochemical changes of the medium such as oxidation-reduction potential, osmotic pressure, and pH can inhibit growth. Thus species of *Aspergillus* can acidify the medium or make it unsuitable for the growth of other organisms. Organisms often compete for a valid nutrient source. Sometimes antibiotic substances are produced.

In some cases of antagonism these mechanisms probably occur together or successively. On artificial media the chief causes are competition for nutrients and the production of antibiotic substances.

The inhibitory action of culture filtrates can be explained by physical changes in the medium during growth and the production of antibiotics. This last point has been the subject of much study ever since the discovery of penicillin. By isolation and purification of compounds in filtrates, antibiotic substances have been obtained which are used in fighting infections in man and animals. But the antibiotics are active against plant pathogens as well.

B. Microbial Antagonisms in Soil

Phytopathologists have neglected for a long time the role of the microbial population of the soil, considering only the parasite and the disease.

Sanford (1926) and Millard and Taylor (1927) began the study of the relations existing between soil microorganisms and the pathogen host complex. Sanford proposed that the beneficent action of certain green-manuring crops in reducing the common scab of potato is indirect and results from the effects of bacteria antagonistic to *Streptomyces scabies*. Millard and Taylor (1927) then suggest that under certain conditions attack by *S. scabies* was reduced by an antagonist, *S. praecox*.

Recent study of biological activity in the soil has resolved paradoxical observations and experimental results, and has suggested the possibility of fighting certain diseases by biological means.

1. *Antagonists Isolated from the Soil.* The microbial population of the soil is abundant. According to Pochon and deBarjac (1958) the number of actinomycetes varies between 100,000 and 36,000,000 per gram in most soil. Bacteria and fungi are also very abundant.

Alexopoulos, studying the antagonistic properties of 80 strains of actinomycetes from the soil, used *Collectrichum gleosporioides* as a test organism and found that 17% of actinomycetes were strongly inhibitory, 39% were slightly inhibitory, and 44% had no action.

In Louisiana, where sugarcane rot occurs, the frequency of actinomycetes antagonistic to the pathogens was studied. From 18% to 31% of the actinomycetes isolated were antagonistic to one of the casual organisms, *Phythium arrhenomanes, Pythium ultimum*, and *Rhizoctonia solani*. Some actinomycetes were antagonistic to all three fungi; others only to one. About 15% of isolated fungi, chiefly species of *Penicillium, Aspergillus*, and *Spicaria*, were antagonistic *Pythium arrhenomanes*. Only 3% of the isolated bacteria were antagonistisc to this fungus. In certain soils where the Panama disease caused by *Fusarium oxysporum* f. *cubense* prevails Meredith (1944) classified the organisms occurring there in three groups: those which apparently increase the pathogen, those having no effect on its growth, and those which are antagonistic to it. Among the 1020 organisms isolated and tested *in vitro*, 66 showed a slight antagonistic action, 39 a middle antagonistic action, and 17 were strongly antagonistic to the pathogen. The distribution of antagonists varied widely according to the different samples of the soil. Among the many antagonistic bacteria was *B. subtilis*, and the most frequently occurring antagonistics fungi were species of *Penicillium, Aspergillus, Trichoderma*, and *Trichothecium*.

Organisms antagonistic to *Ophiobolus graminis* are also common. Thus, Broadfoot (1933) isolated from the soil 66 bacteria and fungi, 15 of which were antagonistic, and Sanford and Broadfoot (1931) noted 36 species antagonistic to *O. graminis*. Some microorganisms, such as *Rhizopus* sp., stimulated this pathogen.

Antagonists in the soil of *Streptomyces scabies, Pythium graminicola*, and *Colletotrichum linicola* have also been reported.

Thus, many fungi, bacteria, and actinomycetes in the soil are antagonistic *in vitro* to pathogens. One wonders what these microorganisms do in the soil and whether they are useful in fighting certain plant diseases.

2. *Action of Antagonists in the Soil on Plant Diseases*. Experiments *in vitro* demonstrate that microbial antagonists can decrease the virulence of phytopathogenic agents and protect the plant from the infection. Sunflower seedlings in Erlenmeyer flasks have been protected from attacks of *Sclerotinia sclerotiorum* by introducing a strain of an antagonistic actinomycete.

Jaarsveld (1942) cultivated Chinese cabbage seedlings in test tubes on sterile media. They were then inoculated with *Rhizoctonia solani*, plus one or several antagonists. Clear antagonistic actions have been obtained with two strains of *Trichoderma lignorum*, followed by *Pyronoma confluence, Cylindrocarpon didymum, Penicillium expansum, Cladosporium herbarium*, and *Absidia spinosa*. The effect of several antagonists seems additive. However, *C. didymum* decreased the antagonism of the others. Disease levels were reduced when experiments were conducted in a previously sterilized soil, but in natural soils results have often been disappointing.

Let us examine some experiments on several types of pathogens and the diseases that they cause.

a. Phytophthora and Pythium. Hartley (1921) several years ago investigated damping-off in the seed bed, caused by *Pythium debaryanum.* The seeds were sown in pots in a previously sterilized soils and then inoculated with the pathogen and various organisms, including *Trichoderma koningi, Phoma* spp., *Chaetomium* sp., *Rhizopus nigricans, Trichothecium roseum, Aspergillus* spp., *Penicillium* spp., *Bacterium* sp., and unidentified fungi. In all cases he emergence of seedlings was increased over the check samples. However, these saprophytes were ineffective in nonsterilized soil.

In the case of the damping-off of tomato seedlings, caused by *Phytophthora parasitica* and *P. cryptogea*, disease levels were reduced in sterile soils when *Aspergillus clavatus* or *Penicillium clavatum* in organic suitable medium were introduced before the pathogen. Glucose improved their effect.

In greenhouse or hotbeds *Trichoderma koningi* decreased the attacks of *P. parasitica* on tomato, and in previously sterilized soils *Pythium* on cucumber seedlings were reduced by a species of *Trichoderma* of a *Pythium* on sugarcane seedings, by actinomycetes isolated from the soil of a *Pythium* on alfalfa, by *Pullularia pullulans* or by *Penicillium expansum*; and of *Pythium arrhenomanes* on wheat roots, by several actinomycetes and fungi belonging to the genera *Spicaria, Penicillium,* and *Aspergillus.*

Many results have been negative, however. Certain organisms, having strongly antagonistic properties *in vitro,* were without action in the soil.

Physical and chemical properties of the soil play an important role. Thus, Wright (1956b) prevented the attacks of *Pythium* on white mustard seedlings by adding *Trichoderma viride, Penicillium nigricans, P. frequentans,* and *P. godlewski* in a lime fertilized soil, while results were negative in acidified soil.

In natural, unsterilized soils, positive results are still more uncommon. Thus in a sterile soil to which 1% glucose had been added *Trichoderma lignorum* or *Streptomyces* sp. considerably reduced the damping-off of alfalfa by *pythium debaryanum* and *P. ultimum.* But in natural soil the results were generally negative, except when a large, antagonistic inoculum was used.

b. Rhizoctonia solani. The reduction of attacks of *Rhizoctonia solani* by *Trichoderma viride* has often been noted. Although *Trichoderma* can parasitize hyphae of *Rhizoctonia*, it also is antagonistic. Some strains of *Trichoderma* produce gliotoxin or viridin.

In sterile soils the addition of *Trichoderma viride* decreases the attacks of Rhizoctonia on cucumber seedlings on peas and on lettuce plants. In natural soils, Cordon and Haenseler (1939) have obtained good results on peas and cucumbers. On lettuce plants Wood (1951) demonstrated inhibition of the pathogen at first, but the effect rapidly disappears. *Bacillus simplex, Bacillus subtilis*, and *Papulospora stoveri* showed themselves antagonistic to the *Rhizoctonia solani* in soil.

c. Phymatotrichum omnivorum. It has been possible to modify the flora in contact with roots by direct inoculation. *Aspergillus luchiensis, Penicillium luteum*, and *Trichoderma lignorum* are all antagonistic to *Phymatotrichum omnivorum* in the laboratory. They can be established and apparently increase in the rhizosphere of the cotton roots in experimental plots. All of these organisms were recovered in greater quantity than in check soils at a distance from the point of introduction. Inconclusive results have been obtained with other antagonistic organisms.

d. Streptomyces scabies. Fungi such as *Trichoderma* are antagonistic to *S. scabies in vitro*. The intensity of potato scab has been decreased by applying this fungus in the furrow around the growing tubers. But even in sterile soils other antagonistic microorganisms (*Penicillium*, bacteria, actinomycetes) had no effect.

e. Ophiobolus graminis. O. graminis decreased in virulence in a sterile soil that then was recolonized by a saprophytic flora. Here the disappearance of the pathogen was sometimes more rapid than in a natural soil, probably because of the growth of strongly antagonistic organisms.

f. Fusarium culmorum and Helminthosporium sativum. In greenhouse soils *Trichoderma lignorum* and *Pyronema confluence* reduced the pathogenicity of *F. culmorum*. When a soil is inoculated with *H. sativum* or *F culmorum*, the virulence of these fungi is decreased by degrees, but much more quickly in a natural than in a sterilized soil, thus suggesting the role of the soil microflora. Further, *h. sativum* cannot colonize unsterilized corn stubbles, whereas it easily establishes itself on this substratum after sterilization.

g. On Different Pathogenic Agents. In experiments in sterile soils species of the genera *Rhizopus, Penicillium*, and *Fusarium* reduced attacks by *Cephalothecium roseum*. In sterilized greenhouse soil different fungi and bacteria have reduced pink root rot (*F. culmorum*) and the bacterial fading of carnation (*P. caryophylli*). *Fusarium udum*, which causes the fading of the gray pea (*Cajanus cajan*) was more virulent in sterile than in natural soil. Several antagonistic organisms were isolated from the natural soil. A species of *Chaetomium* added to a sterilized soil, then infested by *Fusarium lini*, reduced the intensity of the flax wilt. In pots in a sterile soil inoculated with antagonistic bacteria the infection of corn seedlings by *F. graminearum (Gibberella saubinetii)* was prevented. Similarly the inoculation of unsterilized soil on which diseased flax had previously grown reduced the infection by *F. lini* of a second crop of flax seedlings from 26.6% for the tests to 9.5%.

3. *Factors That Modify the Biological Equilibrium and Act on the Disease*. Many factors favor an antagonistic role for microorganisms. We have long known that certain cultural practices alter the attacks of various pathogens, sometimes favorably and sometimes not. Fallow land aids the development of soil saprophytes. But to know whether a cultural practice acts on the virulence of the pathogen or on the resistance of the host directly or whether the effect is indirect by changing the soil-biological equilibrium is not easy. Various studies bear on this point.

The number of actinomycetes in a soil and their relative abundance in relation to the microflora as a whole can be considerably influenced by the crop plants presents, as Lockhead (1940) has shown. Indeed, the roots secrete soluble compound or tissue losses: losses from the root cap and from epidermal and cortical cells. Also, the plants secrete toxic substances into the soil that inhibit certain microorganisms. It is small wonder, then, that the microflora of the rhizosphere varies considerably with the species of plant—even with varieties—and also with their development period.

The action of green manuring on the disease development in relation to potato common scab, caused by *Streptomyces scabies*, and the role of certain microbial antagonists, especially to *S. praecox*, has already been discussed. green manuring with soybean in pots of naturally infested soil has reduced disease, while rye and clover had no effect. Soybean residues considerably increased the population of all soil organisms, especially soil fungi, whereas rye acted only on the bacterial population.

If a soil is poor in organic matter, green manuring makes available quickly the elements favoring development of antagonists. Green manuring can also modify the pH, and some failures have occurred when the soil pH was unfavorable to antagonists.

The influence of green manuring and of amendments and fertilizers on development of the root rot of cereals, caused by *Ophiobolus graminis*, has been extensively studied. This organism grows in two phases: a parasite phase on the growing plants and a saprophytic phase on the roots and stubble. In unsterilized soils its disappearance varies with soil type and with temperature, moisture, and soil fertility.

During the parasitic phase it is possible to reduce the disease by increasing the amounts of assimilable P, nitrate N, organic amendments, oligo-elements, etc. All of these treatments increase the soil microflora, as shown by increased CO_2 production by soils after their addition. In the saprophytic phase the microflora is primarily responsible for the disappearance of the pathogen.

O. graminis is not continuously a saprophyte in the soil. It disappears rapidly as the crop residue is decomposed. It is ill-suited for competing with the soil microflora. It is a poor thing, a guest in an inhospitable society. Its disappearance depends on the soil type and medium conditions. Stubble and material rich in carbohydrates and poor in nitrogen (starch, raygrass flour) hasten its disappearance. A clover crop grown on stubble has the same effect. In this case the action is double. The growth of antagonists is increased, and competition between the crop and the pathogen for the assimilable nitrogen is assured.

In the case of the cotton root rot, caused by *Phymatotrichum omnivorum*, many authors have shown that the addition of organic manures reduces the intensity of the disease in infested soils. The microbiological activity is much more intensive (high release of CO_2) in soils to which such manures have been applied for a long time. The incorporation of green manures in the soil (sorghum, chickpea) also has a good effect. Bacteria and actinomycetes increase considerably, the fungi only a little. Farm yard manure and chopped sorghum also inhibit the growth of *P. omnivorum* in the soil.

Nitrate nitrogen and maize flour reduce the disease levels of *Rhizoctonia solani* and also its persistence in the soil, while sugars, lime, magnesium sulfate, and sulfur favor the disease. Maize flour is especially effective when new antagonists are added to soil. Other amendments give good results in natural soils, but do not act in sterilized soils, thus demonstrating their influence on the soil microflora. Damping-off of alfalfa, caused by *Pythium ultimum* and *P. debaryanum*, is reduced by applying 1% of oat straw and antagonists to a natural infested soil.

Soil temperature influences both the soil microflora and pathogens. Thus-damping-off of alfalfa, reduced by applications of oat straw at low temperatures, was hardly affected at high temperatures. The action of

Trichoderma sp. and *Penicillium vermiculatum* against *Rhizoctonia solani* is much more noticeable at 28° C. than at 18° C.

The pH of the soil affects not only the pathogen but also the action of antagonists. The common scab of potato, caused by *Streptomyces scabies*, and the root rot of wheat, caused by *Ophiobolus graminis*, are frequent in certain limed soils of recently reclaimed health. A biological lack of balance, unfavorable to the antagonists, resulted from the liming. Furthermore, pH influences the action of *Trichoderma viride*.

The addition of a suspension of *Trichoderma* spores in a boggy soil was effective in reducing the attacks of *Rhizoctonia solani* or *Citrus*. The amount of disease reduction was correlated with pH, being excellent for the low pH of 4.5, moderate for a medium pH of 5.7 to 6.1 and inactive at pH 7.0. Soil acidification with aluminum sulfate or sulfuric acid has also been recommended by Wiant (1929) for the control of *Rhizoctonia solani* on forest tree seedlings.

Trichoderma viride is also a natural antagonist of *Fomes annosus* and prevents the establishment of this pathogen on stumps of tree in acid soils. *F. annosus* does not appear in a natural acid soil. However, in the same soil after sterilization *Fomes* becomes established unless it is stopped by introducing *Trichoderma*. The importance of the disease in alkaline soils has been attributed to the lack of *Trichoderma*.

Chemical treatments can also favor antagonistic microorganisms. Carbon disulfide is especially active in reducing the attacks of *Armillaria mellea* on *Citrus*. The pathogen, established on the roots or the trunk of a host plant, is protected by a stratum of pseudosclerotia that permit it to survive in the soil for several years and to resist the attack of soil organisms, especially *Trichoderma viride*. After fumigation *T. viride* is almost always found in the roots of *Citrus* in which the pathogen has been destroyed.

At first, it was thought that *T. viride* resists fumigation and is the first in colonizing the dead mycelium of the pathogen. However, the pathogen does not grow in mixed culture with the antagonist. Added to soils. *T. viride* destroys the pathogen by itself. Moreover, the carbon disulfide is inactive against the pathogen in soil unless the antagonist is present.

Thus, in natural soils *T. viride* is not dominant enough to destroy the *Armillaria*. But after partial sterilization of the soil, *Trichoderma* becomes dominant and colonizes the pathogen.

On occasion 2, 4-D may have similar effects. Thus, Warren et. al. (1951) have reported the prevalence in the soil of an actinomycete after treatment of tomato plants with 2, 4-D. This microorganism on agar medium is antagonistic to many fungi, such as *Rhizoctonia solani, Sclerotinia cinerea, Sclerotium rolfsii*, etc.

4. *Mechanisms of Action of Antagonists*. Different mechanisms of action for antagonism have been noted above. Some postulated mechanisms of antagonism involve the large increase in microflora, following the adoption of a cultural practice, such as the application of a green manure. Large amounts of CO_2, which may affect the pathogen, are produced in this process.

Blair (1943) has suggested such a mechanism for the disappearance of *Rhizoctonia solani*. Alternatively, an increase in the micro-floral activity can lead to the loss of certain elements, particularly of nitrogen.

Sanford (1926) showed that certain bacteria, when cultured in solution, make the solution too acid for the germination of *S. scabies*. Thus the growth of these bacteria in the soil can lead to a modification of the pH, unfavorable to the pathogen.

But very many antagonistic organisms can produce antibiotic substances in culture on artificial media. This raises the problem of the production and role of antibiotics in the soil.

5. *Production and Role of Antibiotics in the Soil*. Most of the known antibiotics are produced by soil microorganisms. We may ask if the soil is a medium suitable for antibiotic production and whether such substances are sufficiently stable in the soil to have a biological role there.

The relations of soil microorganisms in respect to antibiotic production have been studied by Jefferys et. al. (1953) in a sandy and acid podsol. They isolated 65 species, half of which produced antibiotics. The most widely distributed species included the most important percentage of producers. Some common and nonproducing species are comparatively resistant to the antibiotic substances coming from other species. These facts suggest that the production of antibiotics by soil microorganisms may have ecologic significance.

The first experimental work on the production of antibiotic substances in the soil was reported by Grossbard (1948). The author showed that in a sterilized soil enriched with glucose, wheat straw, or beet pulp, *Penicillium patulum* produces an antibacterial substance, patulin. Later studies showed that patulin is produced by other species of *Penicillium* and by 2 species of *Aspergillus* as well when they grow in sterilized soil, enriched with carbohydrates or other assimilable organic substances.

The production of patulin is strongly decreased when other microorganisms are present. Under these conditions the soil must be heavily inoculated with *P. patulum* if antibiotic activity is to appear during the following days. There the production is improved by such organic materials as flour, glucose, and corn steep liquor.

Chloromycetin is produced by *Streptomyces Venezuelae* and actidione by *S. griseus* is sterilized soil amended with various organic elements.

Wright and Grove, (1957) has especially studied antibiotic production by common fungi in the soil. She obtained gliotoxin from a strain of *Trichoderma viride* in two types of sterilized soils (podsol at pH 3.9 and garden soil at pH 6.3), receiving an organic amendment. When no organic matter was added, gliotoxin production was appreciable only in the podsol. Acidification of garden soil favored antibiotic production. In unsterile soils no antibiotic was produced in the garden soil amended with corn flour, but in a podsol enriched with corn flour production was appreciably increased.

For the production of griseofulvin by *P. nigricans* the main factors were sterilization and addition of organic nutrients. But here fungus growth and antibiotic production were better in the garden soil than the podsol. Griseofulvin production was decreased by inoculation of the soil with various microorganisms.

The production of antibiotics in the soil can be influenced by many factors, especially the soil type, its pH, the sterilization, and the addition of organic elements. The beneficial effect of sterilization may eliminate competing organisms and may also increased the availability of nitrogenous organic compounds.

One wonders if the production of antibiotic substances is homogeneous in a soil or if its importance is only locally important near the roots. To answer this question, Wright (1956) studied the production of gliotoxin by *T. viride* in and around straws sunk into the soil. The production was more important in the straw than in the soil around it. The pH of the soil, and especially the pH of the nutrient substratum, had a large effect. The sterilizing of the soil or the straw also played a role.

These studied demonstrate that antibiotics can be produced in the soil at least under some conditions. Presumably antibiotic production can be increased when the influencing factors are better known.

The composition of the nutrient medium has a profound effect on the yield of antibiotic. Thus boron stimulates the production of penicillin; manganese, that of streptomycin; iron and zinc, that of patulin.

Some carbohydrates produce a larger yield than others. *Penicillium notatum* prefers lactose in order to produce penicillin *Streptomyces griseus* prefers starch and glucose for streptomycin and mannose for actidione.

Meat extracts, soybean hydrolyzates, and corn steep liquor provide the amino acids necessary for antibiotic production by the organism. Some compounds serve as precursors, containing groupings found in the molecule of the antibiotic and thus stimulate its production. Examples are phenylacetic acid, β-crotonic acid, glycocoll, and folic acid for penicillin; inositol for streptomycin; tyrosine, phenylalanine, and tryptophan for chloromycetin.

Antibiotics vary in stability in the soil. Gottlieb and Siminoff (1951) noted that streptomycin is so strongly absorbed by clay in the soil that it loses its activity Pramer and Starkey (1951) reported that it disappears more rapidly in unsterile than in sterile soil. Thus in addition to adsorption, degradation by soil microorganism takes place. This was confirmed by Jefferys (1952), who showed the influence of the soil type and its pH. In some cases, when the adsorptive capacity of the soil is saturated, activity can persist when high concentrations are employed, but one has difficulty in seeing how streptomycin can increase in a natural soil and play an appreciable biological role.

Chloromycetin is produced in the soil by *S. venezuelae*. Gottlieb and Siminoff (1952) showed that this neutral antibiotic is not absorbed by clay, and activity persists in sterile soil. However, it is less toxic than in a nutrient medium, the soil having a protecting effect on the sensitive germs. In pure culture various bacteria detoxify chloromycetin by hydrolyzing the molecule or reducing the nitro group. Presumably, microorganisms can have the same effect in soils. In fact chloromycetin disappears in unsterile soils.

Clavacin is another antibiotic that is not readily adsorbed by the clays of the soil. It is stable for a long time in a sterile soil, but the microflora in a normal soil degrade it. The microbial population increases in a soil containing clavacin. Gottlieb et. al. (1952) reported the development of fungus strains resistant to this antibiotic. Differences depending on the soil type have been observed.

Penicillin is detoxified by many microorganisms that secrete penicillinase. In some neutral soils, however, it can maintain its activity sufficiently long to be biologically effective.

Jefferys (1952) studied the stability of ten antibiotics in podsolic soils and in neutral garden soil. Eight of them were produced by fungi isolated from podsols. Some were more stable than others, the stability varying from one soil to another. Four types of inhibition were determined: (1) The natural pH of certain soils is responsible for the instability of albidine, frequentin, gliotoxin, penicillin, and viridin. (2) the soil microorganisms detoxify griseofulvin, mycophenolic acid, and patulin. (3) The soil especially absorbs streptomycin. (4) Inhibition of toxicity occurs in a poorly understood manner, probably of chemical nature. Examples are gladiolic acid, penicillin, and streptomycin. These antibiotics were more stable in acid podsolic soils than in the neutral garden soils. In some soils the stability was sufficient to permit a biological effect when they are produced.

Gottlieb and Siminoff (1952) divide the antibiotics into three groups: the basic antibiotics which are widely inactivated, e.g., streptomycin and streptothrycin; the neutral antibiotics which are slightly inhibited but which maintain a relatively high activity in the soil, e.g., chloromycetin and actidione; and the acid antibiotics, the activity of which is intermediate between the basic and neutral substances, e.g., clavacin.

6. *Antibiotic Action*. In the soil the stable antibiotics that are produced can act directly on pathogenic microorganisms to inhibit their growth. Action of antibiotics *in vitro* includes preventing growth and the deformation—and sometimes the lysis—of vegetative organs. Some antibiotics can also be absorbed by the plant root and translocated to different plant organs without losing their activity. Such substances are called "systemic." A classic example is streptomycin. Plants differ in this respect, however. For example, penicillin is absorbed by bean seedlings, but not by cucumber or maize.

On the other hand, high activity can be found in aerial plant organs only if the antibiotic concentration in contact with the roots is high. Because streptomycin is rapidly inactivated in the soil, it cannot have a systemic role in nature. Chloromycetin, on the contrary, is more stable in the soil, and can be absorbed by the plant if present in sufficient quantity. However, chloromycetin is probably no produced in sufficient quantities in the soil to play a significant biological role in the plant.

The substances produced in the soil by microorganisms do not always play a favorable role. According to Agnihothrudu (1954), *Fusarium oxysporum* f. *vasinfectum* secretes fusaric acid into the soil, and this can be absorbed by roots and damage the plant as it moves through the tissues. The "tiredness"of soils seems to be due in part to the secretion of toxins by fungi such as *Mucor, Alternaria*, or *Cladosporium herbarum*, which live on fragments of diseased plants. Thus, the toxins aid the action of the pathogen.

7. *Discussion: the Biological Struggle among Soil Organisms.* Two aspects of the biological struggle among soil microorganisms can be considered: (1) the specific contribution of antagonistic microorganisms in the soil; and (2) the ways in which the antagonistic functions can be favored.

The first of these can be illustrated simply after sterilization of the soil by heat. Such a soil can be recolonized quickly by antagonistic saprophytes which impede establishment of pathogens. For successful antagonism, favorable conditions must be provided.

Antagonistic organisms often fail to become established in natural soils. In competition with the natural flora, the new organism is often eliminated. However, certain culture methods or chemical treatments sometimes are successful, e.g., fumigations with carbon disulfide, which favors growth of *Trichoderma*.

Gregory et. al. (1952) have shown that the use of antagonists in order to fight the damping-off of leguminous plants may be possible. However, these antagonistic organisms must be inactive against nodule bacteria (*Rhizobium*), which makes the problem complex but not insoluble. Thus an actinomycetes was found which was very active on *Pythium* without affecting the establishment of *Rhizobium* on alfalfa seedings.

We have seen that antibiotics can be produced in the soil. However, when an organism that produces antibiotics is introduced in a soil, it runs the risk of being quickly inhibited. Even if antibiotic production occurs for a short time only this can suffice to control a disease such as damping-off, to which only the seedling plant is susceptible. In this case the antagonist permits the plant to grow through the decisive period of susceptibility, after which it is resistant. Possibly the period of susceptibility can be shortened by increasing the rate of growth of the plant by high temperatures, for example.

Favoring the growth of naturally established antagonists in the soil is more likely to be successful than introducing new ones. The judicious utilization of certain green manures, amendments, and various fertilizing elements will often give positive results. More studies must be made (1) on the growth conditions of antagonists and the production conditions for antibiotic substances, (2) on the stability of these antibiotics and their action mechanism, (3) on the virulence of the pathogen, and (4) on the host susceptibility. These studies, along with some good luck, should bring more precision and success in the biological control of soil-borne pathogens.

C. Antagonism and Competition at the Time of Sowing the Seeds

The normal saprophytic flora of the surface of seeds limits the attack of certain pathogens. Simmonds (1947) reported that the percentage of corn seedlings attacked by *Helminthosporium sativum* was decreased by incubating the seeds for 24 hours in a humid room before insulation. Probably the moisture favored the growth of antagonistic saprophytes on the integuments of the seed.

Ledingham et. al. (1949) and Sallans et. al. (1949) demonstrated that when chemical treatment removed the surface flora from seeds, there was an increase in attack by *H. sativum*. Formaldehyde or lactic acid acted in this way. Thus the seedlings of treated seed which was then inoculated were more frequently diseased than were untreated seed, especially in sterile soil.

This raises a problem about seed disinfection. When the superficial saprophytes are eliminated, pathogens in the integument can grow without hindrance.

In pure culture many organisms which are antagonistic to pathogens have been isolated from the surface of the seed integuments. Attempts have been made to utilize these or other organisms in flighting pathogens that attack seeds.

Novogrudski (1937) recommended bacterial inoculation of seeds to combat *Fusarium lini* and *Colletotrichum linicola* on flax. With various bacteria he was able to decrease the percentage of diseased plants. Beresova and Nauomova (1939) obtained similar results. The attack of *Fusarium graminearum* was reduced by treating corn seed with cultures of *Pseudomonas* or *Achromobacter*. Similarly, Krasilnikov and Raznitzyna (1946), by using similar microorganisms, controlled *Fusarium* on seeds of *Pinus sylvestris*.

By using such species of *Chaetomium* as *C. cochlioides*, Tveit and Wood (1955) obtained as good control of the blight of barley seedlings, caused by *Fusarium nivale*, as when organo-mercurials were used. Apparently the antagonist was able to establish itself and to persist for some time in the soil.

The inoculation of seeds with antagonists may eliminate not only the seed-borne pathogens but soil microorganisms that parasitize the seedlings. Soil saprophytes sometimes decrease the virulence of seed-borne pathogens, as shown by the differences obtained after sowing in natural soil and in sterilized soil. The antagonism phenomena on seeds are doubtless the same as those described in the case of the soil.

Production of antibiotic substances in the integument of seeds was studied by Wright (1956), who noted that pea seedlings arising from seeds inoculated with spores of *Trichoderma viride* had chlorotic cotyledons, similarly to those observed on seedlings growing in gliotoxin solutions. Thus *T. viride* probably produces gliotoxin on the seed in concentrations sufficient to damage the seedlings. This hypothesis was verified: the concentration of gliotoxin in the integuments of pea seeds inoculated with spores of *T. viride* was appreciable 3 days after germination and reached a maximum about the fifth day. The quantity of gliotoxin was much higher in the integument than in the soil around the seed. Moreover, Wright demonstrated that *Penicillium frequentans* could produce frequentin and *Streptomyces venezuelae*, an unidentified antibacterial substance that seems to differ from chloromycetin.

D. Antagonism on Fruit During the Storage

Cases of antagonism have been noted on organisms attacking fruit in storage. Thus two strains of *Bacillus subtilis* are antagonistic *in vitro* to several pathogens of citrus fruit and to *Penicillium digitatum* especially.

E. Antagonisms on a Level with Aerial Organs

The mixing of an antagonist with the inoculum of a pathogen can reduce he infection. This Bamberg (1931) isolated bacteria which *in vitro* were antagonistic to *Ustilago zeae*. The culture filtrate of these bacteria had no effect on the pathogen. But when the bacteria were mixed with smut inoculum, they reduced the infection rate of corn, and inhibited the germination of chlamydospores. They also seemed to cause disintegration of the galls.

In nature the antagonistic saprohytes seems to play a biological role in relation to wound pathogens or pathogens that establish themselves by first colonizing the dead tissues. By invading these tissues before the pathogen does so, antagonists can prevent infection or at least limit it. Thus *Stereum purpureum* establishes itself easily on freshly cut tissues, but does so with difficulty on 3-month-old lesions that are already colonized by saprophytic microorganisms.

The studies of Wood (1951) demonstrate the importance of microflora on leaf tissue. After attacks of *Botrytis* on lettuce in the field more young plants survived in the hollows than where the ground was flat. In the hollows, Wood reasoned, the dried leaves at the base of the plant were sometimes covered by water which had run from the surrounding soil. This permitted the growth of saprophytes antagonistic to the *Botrytis* on the dead tissues. Wood (1951) analyzed this phenomenon experimentally, and concluded, that: many microorganisms are antagonistic to *Botrytis cinerea* at 25° C. but are less so at lower temperatures and that certain of them prevent the rotting of those lettuce leaves when the antagonist is inoculated prior to or simultaneously with *B. cinerea.*

In a similar way Newhook (1957) isolated organisms antagonistic to *B. cinerea* from lettuce and tomato. Cultures of *Bacillus, Pseudomonas,* and *Chromobacterium* from lettuce leaves were more strongly antagonistic to *B. cinerea* at one temperature than at another. On nutrient agar these bacteria raised the pH so high that the growth of *B. cinerea* an its pectolytic activity were inhibited. However, the antagonism was largely due to antibiotic production. In association with many bacteria *B. cinerea* increased or decreased its sporulation, or its hyphae were distorted, in the field organisms isolated from the soil prevented the establishment of *B. cinerea* on dead tissues of different plants. The antagonism of saprophytes must play an important part in the growth of *B. cinerae.* As dampness occurs, it favors the establishment and growth of the pathogen and the development of antagonistic microorganisms simultaneously.

Can antagonistic saprophytes be used to combat pathogens of aerial organs of plants? The growth of saprophytes is possible either with dead tissue is present for colonization or when a suitable nutrient medium is provided for them on the plant. The protection of the living tissues from pathogens by applying nutrients is a theoretical possibility that has not been crowed with success.

However, attempts to establish an antagonistic saprophyte on dead tissues of the host has given some results. *B. cinerea* attacks the fruit only after becoming established on dead petals. In greenhouse experiments Newhook (1957) applied spores of antagonistic saprophytes of *Botrytis cinerea*, especially *Cladosporium herbarum* and one *Penicillium*, to tomato petals direcdy after the fruit was set. This was completely successful when recently dried petals were treated and only 30% successful when applied to petals that had dried for several days. Thus it is possible to guide the colonization of drying organs. On the other hand, natural colonization occurs on organs that have been dead for some time. If these saprophytes are active antagonists of the pathogen, no inoculation need be done because there is a natural protection. If they are not, it seems very difficult to substitute them with others.

Sometimes other treatments have an indirect effect on the colonization by saprophytes of host organs. Newhook (1957) reports that if growth substances are applied to void the fall of tomato fruit, natural colonization by *Cladosporium herbarum* or *Penicillium* is favored. Then, the infection of *Botrytis* is reduced from about 50% on untreated plants to about 2% on treated plants.

How do these antagonistic saprophytes prevent the attack by *Botrytis cinerea* of lettuce and tomato? Newhook (1957) reported that nearly all the saprophytic fungi isolated from the dry petals of tomato are antagonistic to *Botrytis cinerea*. Apparently these fungi have a common method of inhibition that is less specific than antibiotic production. The inhibitation of growth of *B. cinerea* is not a nutrient competition because its spores do not even begin to germinate when they come in contact with petals invaded by saprophytes. In culture the antagonists also prevent spore germination of *Botrytis*. The possibility of an favorable pH in the tissues, created by the antagonist, must also be eliminated.

Among the naturally occurring saprophytes of dry petals of tomato *Cladosporium herbarum* is a most effective antagonist. Some of these saprophytes are very active against *B. cinerea* on sterile tomato petals. Although *Cladosporium herbarum* is not among the most active, it plays an important role in greenhouse culture as an antagonist to *Botrytis* on tomato. *Cladosporium* is able to grow under drier conditions on dying petals than the other microorganisms are. Therefore it more easily colonizes the dead tissues, and thus more completely protects them against *Botrytis* infection.

What are the effects of fungicide treatments on these antagonists? The fungicide, in destroying antagonistic saprophytes, may weaken the natural protection of the dead tissues. But resistant saprophytes, such as *Penicillium*, presumably still protect the dried petals even after treatment. On the other hand, the antagonistic saprophytes that are fungicide sensitive can also protect the dead tissues that were poorly covered with fungicide.

V. INTERACTIONS IN THE FUNGAL AND BACTERIAL DISEASES

When several pathogens are simultaneously or successively establishing on a plant, several cases can exist:

(*a*) A competition occurs on the living tissues of the host. An example is given by *Tilletia foetida* and *T. caries*, both of which infect wheat. While plants in the field are attacked by either pathogen singly, the simultaneous presence of the two *Tilletia* on a same plant is rarely reported. On eliminates the other. The fact has been experimentally demonstrated by Bamberg et. al. (1947). Seeds were artificially inoculated with *T. foetida* and sown in a soil infested with *T. caries*. The presence of *Tilletia foetida* on the seeds and then on the seedlings very clearly reduced the infection by *T. caries*. Another example is *Penicillium digitatum versus Penicillium italicum* on citrus fruits. When citrus fruits are treated with borate of soda, the first fungus is destroyed and the second then grows freely. In this case, *P. digitatum* prevails because of the density of its mycelium.

(*b*) The damage to the plant, caused by one pathogen, can impede the growth of another. When Botrytis fabae attacks the leaflets of field beans and causes them to dry out, attack by *Uromyces fabae* is impeded.

(*c*) The actions of pathogens complement each other in certain simultaneous infections. One pathogen may aid another in entering the host. In other cases on pathogen may provide growth substances needed by another. In still others one may take part is the destruction of living tissues, thus aiding the attack by another. When a wheat plant is attacked by *Ophiobolus graminis*, bacteria can provide thiamine or biotin, substances which increase the virulence of the pathogen.

Sometimes there is a synergism between bacteria and fungi. For instance, Sabet (1954) has demonstrated synergism between *Bacillus polymyxa* nd *Rhizopus nigricans* on potato slices. *Erwinia carotovora* and *B. polymyxa* are synergistic at 35° C. and nearly indifferent at lower temperatures.

O. graminis is more damaging when wheat seedlings have been submitted for several weeks to various organisms, especially *Cladosporium herbarum*. Probably the toxins of *Cladosporium* increase the susceptibility of the plant. This effect is more marked in light soils having a slight absorbent power, because in these soils the toxins move more readily and reach the roots much more easily. Another pair of pathogens that act synergistically are *Diplodia natalentis* and *Colletotrichum gloeosporioides* on the citrus fruits.

(*d*) Fungi attacking one host in serial order may become dominant over the initial invader. The fungi establish themselves only on dead tissue, often follow and sometimes supplant the primary pathogen. The physiologic condition of the host plant or host organ influence this. Thus, the potato tuber which becomes depleted of its reserves in the spring is less and less suitable for the growth of *Phytophthora infestans* and more so for the invasion by pathogens of old tissues.

Successive infections on fruits have been the most thoroughly studied. Scab attack on apples and pears, caused by *Venturia inaequalis* and *Venturia pirina*, favors the establishment of other pathogens: at first, *Monilia fructigena*, the cause of brown rot, and later, during storage, *Trichothecium roseum* and also *Gloeosporium fructigenum* on apples. The invasion by these fungi into the pulp is made easier by the existence of corky callus tissue in the scab spots. Where these join with the skin of the fruit, imperfections occur that offer an easy entry. *Gloeosporium fructigenum* often attacks apples having rust lesions, caused by *Gymnosporangium clavariaeformae*. *Nectria galligena* invades on the branches of apple trees following their injury by *Venturia inaequalis*, the scab pathogen. *Armillaria mellea* very often attacks wal-nut tree roots following *Phytophthora cinnamomi*.

Complex diseases result from antagonism of pathogens or their competition or their synergism on the plant.

Moreau (1957) shows the different stages of the decline of carnations as follows: (1) weakening of the plant by *Uromyces, Heterosporium*, attack by nematodes or poor nutrient conditions; (2) attack by primary pathogens of declining tissue such as *Verticillium cinerescens* or *Fusarium oxysporum* f. *bulbigenum*; (3) invasion by secondary pathogens such as *Fusarium dianthi, Fusarium roseum, Alternaria dianthi*, and *Rhizoctonia solani*; (4) invasion by cellulolytic fungi such as *Rhizoctonia bataticola, Melanospora* sp., *Chaetomium* sp. and of pectolytic fungi such as *Cladosporium herbarum*; and (5) the interaction of the plant microflora with the soil microflora.

Chevaugeon (1954) has also studied complex diseases of manioc. The chief cause are *Glomerella cingulata, Cercospora henningsii*, and *Cercospora caribae*, which favor the establishment of wound and weakness pathogens, which in turn are followed by saprophytes.

VI COMPLEXES ASSOCIATED WITH NEMATODES

Independently of their specific pathogenic action, the phytoparasitic nematodes frequently are involved in the epidemiology of diseases caused by other organisms. The nematodes have many enemies, which influence their numbers in the soil. The apparent effect of edaphic factors on nematode populations is in reality often a result of their effect on the development of natural enemies of nematodes. These effects have been known for a long time. Much of the older work on this subject dealt with the classic species of tylenchids. Biological complexes have much to do with the economic importance of the tylenchids and dorylaimids that are migratory and ectoparasitic.

A. Hyperparasitism in the Nematodes

The parasites of the free nematodes are better known than those of the phytoparasitic species. However, hyperparasitism is more diverse among the plant parasitic nematodes.

1. *Bacteria*. Although they have been noted many times in nematodes, the bacteria are but infrequently pathogenic. However, Schuurmans et. al. (1938) reported an undetermined bacterium to be frequent in *Anguina filiformis*, which frequently showed a partial castration in consequence. In other cases the bacteria can gradually destroy the nematodes. More often they act as symbionts and play a useful role in the digestion.

2. *Microsporidies*. In 1940 Thorne described a microporidian, *Dubosquia penetrans*, on a member of the Nosematidae family. In the southern United States this organism parasitized 66 and 23% of two types of their host nematode, *Pratylenchus pratensis*. This parasite was found later on in other anguiluls.

3. *Fungi*. Many cases are known of parasitism of phytopathogenic nematodes by fungi belonging to very different groups. In 1934 Rozsypal and Schmidt reported a *Protomycopsts chytridiale* attacking *Heterodera schachtii*. The large and round chlamydospores completely filled the encysted females and the larval content had been destroyed.

In 1877 and 1881 J. Kuhn described a phycomycete, *Tarichum auxiliare*, parasitic on nematodes. A thin mycelium enters the cyst by the anal opening and destroys the embroys; from this mycelium persistent spores are formed, which fill the interior of integuments. This genus is related to *Entomophthora*, several species of which attack nematodes. H.

Goffart (1933) recognized that the eggs and embryos of *Heterodera avenae* could be parasitized by an ascomycete, *Cylindrocarpon radicicola* Wollenweber, known to attack underground parts of various cultivated plants.

Some ascomycetes are able to kill nematodes without a direct contact, perhaps by antibiotic action. Metcalf noted the impossibility of growing *Rhabditis bievispina* in the presence of an *Aspergillus.*

B. Predators of Phytophagous Nematodes

Apart from parasitism, nematodes can be attacked by more or less specialized predators.

1. *Protozoa.* In 1952 Weber and associates noted the existence of an "amoebial" organism acting as a predator of the larvae of *Heterodera rostochiensis* in Holland. Called *Theratomyxa weberi* Zwillenberg, it belongs to the family Vampyrellidae in the order of Proteomyxa. This very elastic rhizopod can attack many small-sized species of nematodes, free or parasitic. Many nematodes can be caught by this organism. Thus, 128 larva of *H. rostochiensis* were once observed in one cyst of digestion. Development of this amoeba occurs under such restricted conditions that it cannot be used to control nematodes. It will not tolerate drying.

This species or another very similar was also found in England attacking the golden nematode and in Canada attacking the beet anguillul *H. schachtii.*

2. *Hyphomycetes.* The Hyphomycetes as nematode predators are numerous and frequently met in nature. They have been much studied because of the unique character of their method of capturing the nematodes.

Forty species of these fungi have been found; some have been known for a long time, but most of them were described and studied by Drechsler (1937). These belong to the genera *Dactylaria, Acrostalagmus, Dactylella,* and *Arthrobotrys* (10 species in each of the last two). Some have buckless or adhesive nets that emit mycelium filaments to perforate the cuticle of the immobilized nematode; others have adhesive buds, peduncles, conidia or adhesive hyphae. The most curious traps are formed by constrictor rings, the cells of which suddenly become turgid, thus contracting, when a nematode penetrates into them, e.g., *Dactylella bambicodes.*

Different attempts at practical utilization of these natural enemies have been made, first against the Meloidogynes which attack pineapple in Hawaii, then in France by Roubaud and Deschiens (1939) especially against nematodes parasitic on animals, and recently in England by Duddington in relation to the golden nematode of the potato. Some results have been encouraging, but we are far from using these fungi to control nematodes in agriculture.

3. *Other Nematodes*. Many nematodes, especially those of the Monochidae, are carnivorous and eat other nematodes. They have an enormous chitineous buccal cavity provided with teeth.

In 1927 Thorne showed that huge population of these nematodes occur, especially of *Mononchus papillatus*. The numbers vary widely from one year to another because this species is often parasitized by sporozoans, which destroy them, thus limiting their role as biological control agents.

Other nematodes are also nematophagous. Thus, dorylaimids are frequently found inside *Heterodera* cysts; the eggs are empty.

4. *Other Nematophages*. Although we often lack precise experiments, it seems that many enchytreides, various Acari, and other soil arthropods attack nematodes that they meet.

C. Action of Root Secretions on Nematodes

The action of the flora on the development of the eelworm population can be important because this population is frequently limited by the amount of food in the environment. Obviously the growth of specialized nematodes that can live only on some hosts depends on the frequency with which the host occurs, either as a crop or in the self-sown flora. But to many nematodes, especially Heterodera, the action of plants is more complex. Here the plant host serves not merely as a host, but also induces the larvae to hatch from cysts in the soil, and attracts the larvae.

Sometimes plants exert this action on nematodes which cannot grow in their tissues. Thus in Holland, Hijner (1951) tried to use the wild beet *Beta patellaris* in a practical way because it induces hatching of cysts of *H. schachtii* without permitting the larvae to grow on its roots. The larvae then die for want of a host.

The chemical nature of these hatching factors is generally difficult to determine and not yet known. However, the work of Calam et. al. (1949) and of Marrian et. al. (1949) in Great Britain has resulted in the isolation of eclepic acid which is secreted by the potato and stimulates the hatching of the golden nematode *H. rostochiensts*. These authors showed that anhydrotetronic acid has the same property.

In contrast, some plants inhibit the hatching of larvae from eggs even when they are later exposed to hatching factors. The researches of Triffet (1934), Goffart (1934), and Franklin (1937) have described the action of root secretions of many plants on *H. rostochiensis*. Thus, most leguiminous plants, white mustard (*Sinapis alba*), maized, various fodder grasses, and other grasses strongly reduce the viability of cysts. Franklin (1937) reports a reduction of 73% by exposure of cysts to *Poa trivialis* and *Poa pratensis*.

Other plants have a direct toxic action on many nematodes such as *Tagetes*, which practically eliminates certain soil species. Their nematocidal effect has recently been demonstrated. The toxic substance is terthienyl, which is active on certain tylenchids—especially larvae of *Heterodera*— at doses as low as 0.1 p.p.m. The dosage required is low for other nematodes as well, e.g., *Ditylenchus dipsaci*, *Anguina tritici*, and *Pratylenchus* sp. However, the substance has low activity for the neighboring genera *Meloidogyne* and *Hoplolaimus*.

D. Interaction between Nematodes and Other Pathogenic Agents in Complex Diseases

1. *Cryptogamic Diseases.* As early as 1892 Atkinson noted the simultaneous presence in cotton of the root knot nematode and fusarial wilt. The combined action caused much more severe damage than was caused by either pathogen alone. The use of wilt-resistant varieties has permitted evaluation of the damage done by different nematodes belonging to the genera *Meloidogyne* (causing root knot), *Pratylenchus* (the meadow nematode), and *Belenolamus* (sting nematode). Holdeman and Graham (1953) have demonstrated experimentally the role of the latter in increasing the aggressiveness of wilt foı resistant varieties.

The different nematodes affect their host in characteristic ways. Some destroy the root endings, thus providing an entry for fungi. Others inject toxic salivary secretions that cause histologic anomalies in the host that in turn play a similar role in relation to fungi.

More complex interactions can be involved. A nematode may weaken the host generally or cause a specific nutrient imbalance, so that the host becomes more susceptible to fungal attack. An example in cotton is the high potassium required for growth of *Meloidogyne*, which leads to potassium deficiency of the host and greater susceptibility to cotton wilt.

Nematodes and soil fungi are frequently associated in disease. Examples are the association between *Meloidogyne* sp. and black shank of tobacco, caused by *Phytophthora parasitica* var. *nicotina* (Tisdale, 1931), and that between root knot and tomato wilt.

Nematodes may act as vectors for fungi, as shown by Barat (1952) for the decline of pepper plants in Indochina. Thus larvae of *Meloidogyne* were reared aseptically on plants attacked by *Fusarium* or *Pythium*. When these nematodes were allowed to attack healthy plants, these fungi rapidly invaded the roots. When similarly reared larvae were treated with bichloride of mercury before being transferred to healthy plant, however, the nematodes grew normally and the roots were not invaded by fungi.

2. *Bacterial Diseases*. Bacteria and nematodes are sometimes associated in complex diseases. Thus, Crosse and Pitcher (1952) showed that the leafy gall of strawberry is due to the simultaneous presence in the foliage of the bacterium *Corynebacterium fascians* and an *Aphlenchoides*. Alone, neither one produces conspicuous symptoms, but together they produce a disease that leads to monstrous plants. Similar effects have been reported for a disease of wheat, caused by *Corynebacterium tritici*, in association with the nematode *Anguina tritici*. The work of Stewart and Schindler (1956) showed that 4 species of *Meloidogyne* as well as *Helicotylenchus nannus* all accelerate the rate of wilting of carnations attacked by the bacterium *Pseudomonas caryophylli*. Except for *M. incognita*, these nematodes alone produced no such effect. When other nematodes attack bacterially wilted carnations, they do not accelerate wilting.

Lucas and Krusberg (1956) observed the same phenomenon on the bacterial wilt of tobacco, caused by *Xanthomonas solancearum*, and explained it by the nature of the attack of the nematode. *Tylenchorhynchus claytoni*, which has no action on wilt, remains very superficially in the cortex, whereas *Meloidogyne*, which increases the damage from wilt, changes the vascular bundles.

3. *Conclusion*. The nematodes together with soil-borne microbial pathogens play an important role in complex diseases. Also they frequently accentuate damage caused by other pathogens. Because the nematodes sometimes increase the damage and sometimes reduce it, we must be cautious in interpreting the observed facts. For example, some

years ago Lambert and associates (1949) showed that the "*Cephalothecium disease*" of cultivated mushrooms was really due to a *Ditylenchus* that is always associated with a fungus, *Arthrobotrys superba*. Formerly the constant association of the *Arthrobotrys* with this condition misled investigators to think of it as the cause. The fungus does not cause the disease. In fact in reduces the disease by being a predator of the nematode.

The multitude of soil microorganisms leads to many biological equilibria. Linford and Oliveiria (1938) counted no less than 52 predators and parasites of the *Meloidogyne* that attack pineapples in Hawaii. There were 18 fungi, 1 protozoan; 24 nematodes, 6 acaridae, and 3 tardigrads. Many of them also attack one another. These interactions, the role of which can be important, have often been neglected in our thinking and have been destroyed before the onslaught of the classic nematocidal treatments.

VII. INTERFERENCES IN THE VIRUS DISEASE

Interactions between plant viruses and other organisms are little known. For example, potatoes infested by the leaf roll virus are more actively colonized by Doryphors and are quickly eaten. Probably the starch accumulation resulting from virus infection offers a choice food for the Doryphor, which increases its voracity and its proliferation capacity tenfold.

More thorough observations have been made on the relations between cryptogamic and viral diseases. The presence of the viruses X and Y in the potato increases its resistance to late blight, caused by *Phytophthora infestans*. The leaf roll virus has the same effect on the potato plant, yet this virus is considered to increase the severity of the late blight. This effect probably results from the rolling of leaves which retain rain water, thus creating a microclimate favorable to the pathogen.

The case where a microorganism creates an environment favoring a virus disease is also known. For example, beans infected by the rust *Uromyces phaseoli* have a content of viruses 50 times superior to the plants free from rust.

A. Interaction between Viruses in the Host

Two different viruses may infect the same plant, thus producing a complex disease. For example, the tomato filiform disease is caused by virus 1 of cucumber and tobacco mosaic virus. The double streak of tomato and tobacco is caused by virus X of potato and tobacco mosaic virus. In both these cases the complex aggravates the symptoms caused by each virus separately. Together they lead to a necrotic disease, the severity of which depends on the virulence of the virus X strain involved. Therefore, virus X is the determinant, and its concentration probably is modified by the presence of tobacco mosaic virus. A similar effect of one virus on the multiplication rate of another in the same host occurs with virus X and Y of potato when they attack potato, tobacco, or *Nicotiana glutinosa*. The titer of virus X is increased up to 10 times the normal concentration if virus Y is inoculated at the same time or afterwards. If virus Y is inoculated first, the titer is less.

Two unrelated viruses can antagonize one another. One of the most curious cases is that of the severe etch virus, which not only prevents the multiplication of the virus Y of the potato and virus 3 of henbane, but even eliminates them from an infected plant.

Interaction between viruses, leading to a synergy or inhibition, is not difficult to explain. If each of two unrelated viruses affects the troubles of metabolism of the host in a specific way, an effect on their multiplication could be expected, Bawden and Kassanis (1945) have supposed that antagonism probably resulted from a decrease in the host of either the metabolites or enzymatic system that are necessary to the multiplication of virus Y and virus 3 of the henbane.

In complex diseases the appearance of symptoms can be explained by assuming that symptoms arise from physiological aberrancies in the plant. Thus, if one virus instigates modifications too slight to the translated into symptoms, then two viruses acting together can sometimes produce a visible effect. For example, the necrotic lesions in the case of the double streak caused by a combination of virus X and tobacco mosaic virus.

B. Interactions between Strains of the Same Virus

A plant completely infected cannot be reinfected by this same virus or, to be more precise, by the same strain of this same virus. Protection of a plant against reinfection with the same virus is most evident in the case of diseases having a chronic phase. Tobacco ring spot is manifested by a shock phase immediately after inoculation, followed by a chronic phase during which the content of viruses declines and symptoms disappear. Eventually certain organs are free of virus. When the host is inoculated during the chronic phase, no symptoms are obtained; if it is inoculated on a cured organ, shock symptoms appear.

A plant is not only protected against the strain of virus with which it is infected, but also against all other strains of this virus. This type of interference constitutes premunity and is observed only between closely related viruses. It can be seen only when inoculation is done mechanically. Inoculation by insect vectors and by grafting rarely permit this phenomenon to be observed. A host plant must have been inoculated for a few days in order to be protected against reinoculation. When two strains are inoculate into the host simultaneously, they multiply in competition.

Thung (1931) showed evidence that tobacco infected by the common strain of tobacco mosaic virus is protected against inoculation by the white strains. Salaman (1933) protected potatoes infected by a strain of low virulence against the more virulent strains of the virus X. Recently, Kunkel (1955) has shown that the usual strain of aster yellows protects the plant not only against the California strain of the same virus, but also protects the insect vector *Macrosteles fascifrons* against the California strain, and vice versa. The virus of the aster yellows multiplies in the insect vector as well as in the plant. This similarity of protection of both plant and animal hosts permits an explanation of premunition that can be accepted.

1. *Antibody Theory.* According to the antibody theory, the plant as well as the animal is able to produce immunizing diffusible substances neutralizing a virus. This theory, which is open to questions in some areas, has, however, the advantage of explaining the strict specificity of premunity. Indeed, interference is produced only between strains of the same virus. This is the best evidence in favor of the presence of antibodies, provided by the curly top virus on tobacco.

The curly top virus of sugar beet illustrates the degrees of interference possible. Protection between different strains is nonexistent in beets. Inoculated into tobacco the virus produces an acute disease followed by a chronic phase. When a tobacco scion in the chronic phase is grafted onto a tomato, a less serious disease results. If, instead of grafting, the insect vector is used to transmit the virus from tobacco to tomato, an acute disease is obtained. Supposedly, antibodies have moved with the virus in grafting.

2. *Theory of Exhaustion of the Precursor of Virus.* The metabolites necessary for the virus synthesis are completely utilized by the first strain, so that a new infection cannot succeed. When the virus is inocu-

lated into a cell, it multiplies rapidly after a latent period, until a certain concentration is attained, when all precursors are converted to viruses. Thereafter the rate of increase is drastically reduced.

The contrary situation was found to exist by Bawden and Kassanis (1945). A strain of the etch virus of low virulence which, with but slight concentration in the plant at best, protects against a virulent strain, the concentration of which is usually much higher. This type of protection obviously cannot be due to exhaustion of the precursor.

3. *Theory of the Occupied Receivers.* To explain the phenomenon described above, Hutton and Bawden (1950) suggest that unrelated viruses increase on receivers or on specific increasing surfaces. When a virus finds these receivers occupied by the first strain inoculated, they are incapable of duplication.

All the theories recognize that interference between strains of a virus is due to competition. Either the first strain launches the attack on host organs, which will not permit a new inoculation, it exhausts the nutrient reserve of the cell, or it quickly occupies a space. In all these suppositions no direct interaction exists from one virus strain to another strain.